企业级卓越人才培养解决方案"十三五"规划教材

Ionic 项目实战

天津滨海迅腾科技集团有限公司　主编

U0265188

南开大学出版社
天　津

图书在版编目 (CIP) 数据

Ionic 项目实战 / 天津滨海迅腾科技集团有限公司主编. — 天津：南开大学出版社，2018.7
ISBN 978-7-310-05613-2

Ⅰ.①I… Ⅱ.①天… Ⅲ.①移动终端－应用程序－程序设计 Ⅳ.①TN929.53

中国版本图书馆 CIP 数据核字 (2018) 第 131995 号

主　编　常秀岩　郝振波
副主编　李树真　刘　盟　杨海娇

版权所有　侵权必究

南开大学出版社出版发行
出版人：刘运峰
地址：天津市南开区卫津路 94 号　　邮政编码：300071
营销部电话：(022)23508339　23500755
营销部传真：(022)23508542　　邮购部电话：(022)23502200
*
天津午阳印刷有限公司印刷
全国各地新华书店经销
*
2018 年 7 月第 1 版　　2018 年 7 月第 1 次印刷
260×185 毫米　16 开本　20 印张　500 千字
定价：69.00 元

如遇图书印装质量问题，请与本社营销部联系调换，电话：(022)23507125

企业级卓越人才培养解决方案"十三五"规划教材编写委员会

指导专家： 周凤华　教育部职业技术教育中心研究所
　　　　　　李　伟　中国科学院计算技术研究所
　　　　　　张齐勋　北京大学
　　　　　　朱耀庭　南开大学
　　　　　　潘海生　天津大学
　　　　　　董永峰　河北工业大学
　　　　　　邓　蓓　天津中德应用技术大学
　　　　　　许世杰　中国职业技术教育网
　　　　　　郭红旗　天津软件行业协会
　　　　　　周　鹏　天津市工业和信息化委员会教育中心
　　　　　　邵荣强　天津滨海迅腾科技集团有限公司
主任委员： 王新强　天津中德应用技术大学
副主任委员： 张景强　天津职业大学
　　　　　　宋国庆　天津电子信息职业技术学院
　　　　　　闫　坤　天津机电职业技术学院
　　　　　　刘　胜　天津城市职业学院
　　　　　　郭社军　河北交通职业技术学院
　　　　　　刘少坤　河北工业职业技术学院
　　　　　　麻士琦　衡水职业技术学院
　　　　　　尹立云　宣化科技职业学院
　　　　　　廉新宇　唐山工业职业技术学院
　　　　　　张　捷　唐山科技职业技术学院
　　　　　　杜树宇　山东铝业职业学院
　　　　　　张　晖　山东药品食品职业学院
　　　　　　梁菊红　山东轻工职业学院
　　　　　　赵红军　山东工业职业学院
　　　　　　祝瑞玲　山东传媒职业学院
　　　　　　王建国　烟台黄金职业学院

陈章侠	德州职业技术学院
郑开阳	枣庄职业学院
张洪忠	临沂职业学院
常中华	青岛职业技术学院
刘月红	晋中职业技术学院
赵　娟	山西旅游职业学院
陈　炯	山西职业技术学院
陈怀玉	山西经贸职业学院
范文涵	山西财贸职业技术学院
郭长庚	许昌职业技术学院
许国强	湖南有色金属职业技术学院
孙　刚	南京信息职业技术学院
张雅珍	陕西工商职业学院
王国强	甘肃交通职业技术学院
周仲文	四川广播电视大学
杨志超	四川华新现代职业学院
董新民	安徽国际商务职业学院
谭维奇	安庆职业技术学院
张　燕	南开大学出版社

企业级卓越人才培养解决方案简介

企业级卓越人才培养解决方案（以下简称"解决方案"）是面向我国职业教育量身定制的应用型、技术技能型人才培养解决方案，以教育部-滨海迅腾科技集团产学合作协同育人项目为依托，依靠集团研发实力，联合国内职业教育领域相关政策研究机构、行业、企业、职业院校共同研究与实践的科研成果。本解决方案坚持"创新校企融合协同育人，推进校企合作模式改革"的宗旨，消化吸收德国"双元制"应用型人才培养模式，深入践行"基于工作过程"的技术技能型人才培养，设立工程实践创新培养的企业化培养解决方案。在服务国家战略，京津冀教育协同发展、中国制造2025（工业信息化）等领域培养不同层次的技术技能人才，为推进我国实现教育现代化发挥积极作用。

该解决方案由"初、中、高级工程师"三个阶段构成，包含技术技能人才培养方案、专业教程、课程标准、数字资源包（标准课程包、企业项目包）、考评体系、认证体系、教学管理体系、就业管理体系等于一体。采用校企融合、产学融合、师资融合的模式在高校内共建大数据学院、虚拟现实技术学院、电子商务学院、艺术设计学院、互联网学院、软件学院、智慧物流学院、智能制造学院、工程师培养基地的方式，开展"卓越工程师培养计划"，开设系列"卓越工程师班"，"将企业人才需求标准、工作流程、研发项目、考评体系、一线工程师、准职业人才培养体系、企业管理体系引进课堂"，充分发挥校企双方特长，推动校企、校际合作，促进区域优质资源共建共享，实现卓越人才培养目标，达到企业人才培养及招录的标准。本解决方案已在全国近几十所高校开始实施，目前已形成企业、高校、学生三方共赢格局。未来三年将在100所以上高校实施，实现每年培养学生规模达到五万人以上。

天津滨海迅腾科技集团有限公司创建于2008年，是以IT产业为主导的高科技企业集团。集团业务范围已覆盖信息化集成、软件研发、职业教育、电子商务、互联网服务、生物科技、健康产业、日化产业等。集团以产业为背景，与高校共同开展产教融合、校企合作，培养了一批批互联网行业应用型技术人才，并吸纳大批毕业生加入集团，打造了以博士、硕士、企业一线工程师为主导的科研团队。集团先后荣获：天津市"五一"劳动奖状先进集体，天津市政府授予"AAA"级劳动关系和谐企业，天津市"文明单位"，天津市"工人先锋号"，天津市"青年文明号""功勋企业""科技小巨人企业""高科技型领军企业"等近百项荣誉。

前　言

　　Ionic 是一款应用广泛的跨平台移动开发框架,具有开源、性能优越、简单易学等特点。Ionic 最大的亮点是基于主流技术 HTML5、Angular 和 Cordova,极大的节约了开发时间和成本并融入了成熟的前端工程技术。

　　本书详细讲解了 Ionic 跨平台开发框架环境的搭建、项目开发流程、开发步骤及测试部署等知识,内容系统而全面。本书编写的目的是让读者能够把理论与实践完美的结合,从而达到熟练掌握使用 Ionic 快速开发跨平台 APP 的技能。

　　本书主要以项目为基础,由八个项目组成,项目一至项目六主要从"项目结构"→"导航组件"→"Ionic 常用事件"→"极光推送插件"→"加载指示器"→"存储数据"等相关知识进行介绍,从而让读者掌握 Ionic 整体结构和知识技能,能够根据所学知识做出跨平台 APP 界面;项目七主要讲解 Ionic 服务器模拟环境搭建,以 Postman 安装与使用示例、MongoDB 安装与测试、使用 Mongoose 完善数据持久化等进行介绍;项目八对如何生成发布 Android/iOS 平台的应用包进行介绍,使读者能够对跨平台项目进行测试、部署。

　　本书的每个项目都分为学习目标、学习路径、任务描述、任务技能、任务实施、任务总结、英语角、任务习题八个模块来讲解相应的知识点。此结构条理清晰、内容详细,任务实施可以将所学的理论知识充分的应用到实战中。

　　本书由常秀岩、郝振波任主编,李树真、刘盟、杨海娇等任副主编,由常秀岩、郝振波负责统稿,李树真、刘盟、杨海娇负责整体内容的规划与编排。具体分工如下:项目一至项目三由常秀岩、郝振波共同编写,常秀岩负责全面规划;项目四至项目六由李树真、刘盟编写,李树真负责全面规划;项目七至项目八由郝振波、杨海娇编写,杨海娇负责全面规划。

　　本书理论内容简明、扼要、即学即用;实例操作讲解细致,步骤清晰,实现了理论点与实践的结合。不仅讲述了使用 Ionic 框架开发的跨平台应用(Android 和 iOS)打包的完整过程。除此之外还介绍了发布和更新应用的方法,使读者能真正将开发的应用转化为经济效益。

天津滨海迅腾科技集团有限公司
技术研发部

目 录

项目一 "Dancer 时代"登录模块的实现 ·· 1
    学习目标 ··· 1
    学习路径 ··· 1
    任务描述 ··· 1
    任务技能 ··· 3
        技能点 1　跨平台移动开发框架 ··· 3
        技能点 2　Ionic 环境配置 ·· 6
        技能点 3　Ionic 项目结构 ··· 14
    任务实施 ··· 20
    任务总结 ··· 30
    英语角 ·· 31
    任务习题 ··· 31

项目二 "Dancer 时代"首页模块的实现 ·· 32
    学习目标 ··· 32
    学习路径 ··· 32
    任务描述 ··· 32
    任务技能 ··· 33
        技能点 1　Ionic 的 CSS 组件 ·· 33
        技能点 2　导航组件 ·· 43
        技能点 3　Ionic 列表 ·· 49
        技能点 4　Ionic 功能 ·· 54
    任务实施 ··· 61
    任务总结 ··· 73
    英语角 ·· 73
    任务习题 ··· 73

项目三 "Dancer 时代"音频模块的实现 ·· 75
    学习目标 ··· 75
    学习路径 ··· 75
    任务描述 ··· 75
    任务技能 ··· 77
        技能点 1　Ionic 弹出框 ··· 77

　　　　技能点 2　常用事件 ··· 89
　　　　技能点 3　插件简介及使用 ··· 93
　　　　技能点 4　音乐播放 ··· 98
　　任务实施 ·· 103
　　任务总结 ·· 135
　　英语角 ··· 135
　　任务习题 ·· 136

项目四　"Dancer 时代"上传模块的实现 ·· 138

　　学习目标 ·· 138
　　学习路径 ·· 138
　　任务描述 ·· 138
　　任务技能 ·· 139
　　　　技能点 1　Ionic 选择器插件 ·· 139
　　　　技能点 2　访问设备文件插件 ·· 151
　　　　技能点 3　图片预览 ··· 154
　　　　技能点 4　下拉刷新实现 ·· 157
　　　　技能点 5　Ionic 极光推送 ··· 160
　　任务实施 ·· 164
　　任务总结 ·· 174
　　英语角 ··· 175
　　任务习题 ·· 175

项目五　"Dancer 时代"分享模块的实现 ·· 177

　　学习目标 ·· 177
　　学习路径 ·· 177
　　任务描述 ·· 177
　　任务技能 ·· 179
　　　　技能点 1　指纹验证 ·· 179
　　　　技能点 2　加载指示器 ··· 183
　　　　技能点 3　社交分享插件 ·· 186
　　　　技能点 4　地图定位 ·· 188
　　任务实施 ·· 193
　　任务总结 ·· 221
　　英语角 ··· 221
　　任务习题 ·· 221

项目六　"Dancer 时代"我的模块实现 ··· 223

　　学习目标 ·· 223
　　学习路径 ·· 223

任务描述 ·········· 223
　　任务技能 ·········· 224
　　　技能点1　媒体捕获 ·········· 224
　　　技能点2　扫描二维码 ·········· 227
　　　技能点3　语音识别 ·········· 230
　　　技能点4　拨打电话 ·········· 231
　　　技能点5　存储数据 ·········· 233
　　任务实施 ·········· 237
　　任务总结 ·········· 248
　　英语角 ·········· 249
　　任务习题 ·········· 249

项目七　Ionic 服务器模拟环境搭建 ·········· 250

　　学习目标 ·········· 250
　　学习路径 ·········· 250
　　任务描述 ·········· 250
　　任务技能 ·········· 251
　　　技能点1　Postman 安装与使用示例 ·········· 251
　　　技能点2　使用 Express 初始化创建 API 示例 ·········· 256
　　　技能点3　MongoDB 安装与测试 ·········· 261
　　　技能点4　使用 Mongoose 完善数据持久化 ·········· 265
　　任务实施 ·········· 266
　　任务总结 ·········· 277
　　英语角 ·········· 277
　　任务习题 ·········· 278

项目八　"Dancer 时代"发布 ·········· 279

　　学习路径 ·········· 279
　　任务描述 ·········· 279
　　任务技能 ·········· 280
　　　技能点1　生成发布 Android 平台的应用包 ·········· 280
　　　技能点2　生成发布 IOS 平台的应用 ·········· 285
　　任务总结 ·········· 303
　　英语角 ·········· 303
　　任务习题 ·········· 304

项目一 "Dancer 时代"登录模块的实现

通过"Dancer 时代"登录模块的实现,了解登录功能的实现流程,学习 Ionic 框架的优缺点,掌握 Ionic 所需软件的安装及环境的配置,具有安装 Ionic 环境及制作出 Ionic 界面的能力。在任务实现过程中:

- 了解登录功能的实现流程。
- 学习 Ionic 项目的基本结构。
- 掌握 Ionic 项目的创建。
- 具有使用 Ionic 制作界面的能力。

【情境导入】

在这个衣食无忧,科技发达的时代,越来越多的人开始追求精神上的需求,而舞蹈正是满足大众需求的一种文化形式。经过讨论,项目负责人 Richard 和他的开发团队决定开发一款手机软件,让更多的人在学习舞蹈的同时也可以了解舞蹈中蕴含的文化。Richard 等人将该软件命名为:"Dancer 时代"。"Dancer 时代"主要包括六个模块,分别为:登录、首页、音频、上传、分享、我的。本项目主要通过实现"Dancer 时代"的登录功能了解 Ionic 的基础知识和基本项目结构。

【功能描述】

本项目将实现"Dancer 时代"登录界面。
- 使用 Ionic 轮播组件实现轮播图效果。
- 使用路由进行页面的跳转。
- 使用输入框组件实现输入框的样式。
- 使用 ngModel 进行数据的双向绑定。

【基本框架】

基本框架如图 1.1 和图 1.3 所示,通过本项目的学习,能将它们转换成效果图 1.2 和图 1.4。

图 1.1　框架图 1　　　　　　　图 1.2　效果图 1

图 1.3　框架图 2　　　　　　　图 1.4　效果图 2

技能点 1 跨平台移动开发框架

1 跨平台移动开发框架简介

随着移动互联网的飞速发展,跨平台开发已成为移动应用程序必不可少的元素,原生移动应用已不能满足开发的需求,为此国内外的专业开发团队应发展需求推出多种不同跨平台移动开发框架。跨平台移动开发框架是一种在不依赖操作系统和硬件环境的情况下,通过打包工具适配输出,并可以在多种操作系统上运行的开发框架。它能够降低软件开发周期、减少软件维护的费用,提升软件使用时间。

2 跨平台移动开发框架

标准化的语言规范、丰富的开发资源是跨平台移动开发最大的优点,跨平台移动开发框架为其提供方便的打包、发布服务、实用的 API、灵活的扩展机制。使用跨平台移动开发框架具有以下优势:

(1)平台重用代码

跨平台移动开发框架通过极少修改甚至不用修改的情况下可以在另一个操作系统下顺利运行。该框架可以提高开发速度并降低成本。

(2)使用难度低

跨平台移动开发框架采用 HTML5/CSS3/JavaScript 为主的开发语言平台,通过对已有应用模板的定制修改扩展,开发出可用的 APP 应用。

(3)实时更新推送

通过网页形式动态渲染界面的跨平台移动开发框架,辅以动态加载组件进行更新推送。

(4)自带样式

跨平台移动开发框架自身拥有满足移动开发的组件和插件,除此之外,还可自定义组件和插件。

对于开发一款手机 APP,选择跨平台移动开发框架是非常重要的。Mobile Angular UI、Ionic 和 Sencha Touch 是如今发展较为成熟、使用率较高的三款跨平台开发框架。

(1)Mobile Angular UI

Mobile Angular UI 是一款移动 UI 框架,是 Bootstrap 框架的扩展,用于快速构建移动应用程序。Mobile Angular UI 并不包含任何 jQuery 依赖,只是通过一些 Angular 指令创建友好的用户体验。它使用的是 Bootstrap3 的语法和一些特定的移动部件,并且通过添加 CSS 使内容可以响应、具备触摸功能,从而将 Web 应用程序转换为移动版本。Mobile Angular UI 框架具有以下特点：
- iScroll 的可滚动区域；
- 滑出/滑入侧边栏；
- 底部导航条；
- 合理的 Buttons 设计；
- 使用 Grunt 自定义构建工作流；
- 比 Bootstrap3+jQuery+Angular 更轻量级。

（2）Ionic

Ionic 是一个强大的 HTML5 应用程序开发框架，可以通过 Cordova 打包成移动 APP，只需编写一次代码，即可部署到 IOS、Android 等多种移动平台上。Ionic3 使用 TypeScript 进行开发，具有以下优点：
- ➢ 更快的性能；
- ➢ 更清晰的项目结构；
- ➢ 更强大的 CLI；
- ➢ 更友好的页面导航；
- ➢ 更高效的开发体验。

（3）Sencha Touch

Sencha Touch 是 HTML5 手机应用跨平台开发框架，可以兼容 IOS、Android、Blackberry 系统。其拥有良好的用户界面组件和丰富的数据管理，基于 HTML5 和 CSS3 的 Web 标准，全面兼容 Android 和 IOS 设备。Sencha Touch 的优势如下：

※ 基于 Web 标准（HTML5+CSS3+JavaScript）。

整个库在压缩后大约 80 KB，通过禁用一些组件还会使它更小。

※ 良好的触摸事件。在 TouchStart、TouchEnd 等标准事件基础上，增加了一组自定义事件，如 tap、swipe、pinch、rotate 等。

※ 数据集成。提供了强大的数据包，通过 AJAX、JSONP、YQL 等方式绑定到组件模板，写入本地离线存储。

表 1.1 为三种框架对比情况。

表 1.1 跨平台框架对比

	Mobile Angular UI	Ionic	Sencha Touch
项目结构	Mobile Angular UI 框架将界面的 HTML 文件与 JS 文件分别放在两个文件夹中	在 Ionic3 中，每个组件、页面都只专注于做一件事，有专属的一个目录、专属的类（Class）、模板文件（Template）和样式文件（SCSS）。项目结构更加简洁	Sencha Touch 框架将界面的 HTML 文件与 JS 文件分别放在两个文件夹中

续表

	Mobile Angular UI	Ionic	Sencha Touch
项目性能	Mobile Angular UI 集成了 Angular、Bootstrap3、Angular 的 Bootstrap3 指令集、一系列重要的移动端 Bootstrap3 组件等构建了高性能的移动项目	界面跳转流畅 项目运行速度快 项目内存小	Sencha Touch 的性能非常的低，项目里有图片的时候，在 Android 手机上运行速度非常慢
兼容性		支持 IOS6 和 Android4.1 以上的版本	兼容 IOS、Android、Blackberry 系统

3 跨平台移动开发框架选择

在对跨平台移动开发框架进行初步了解后，从项目开发的角度看，使用 Ionic 框架可以有效利用 Angular 的特性，极大地提高 HTML5 应用开发的效率、质量、模块化程度。选择 Ionic 的理由如下：

（1）社区支持服务

Ionic 有专人在社区进行问题解答。当有 Bug 暂未解决时，官方文档网站也会及时明确提示。

（2）性能优势

手机运行流畅，看不出和原生差别。

（3）完整的开发构建工具链

通过使用 NPM、Gulp、Bower 等组成的工具链，能够快速进入 APP 应用的迭代开发测试阶段，节省大量的开发时间。

（4）开发成熟

经过 Ionic 框架开发团队不断改进，Github 上该开源项目的发布总数趋于不变，并且在各个网站上 Ionic 新问题越来越少，目前 Ionic 达到了成熟可控没有 Bug 的程度。

提示 当对跨平台移动开发框架有所了解之后，是否想更详细地知道为什么本书选择介绍 Ionic 框架？扫描二维码，会解决你的困惑！

技能点 2　Ionic 环境配置

在使用 Ionic 开发移动应用之前，需要安装 Node.js、Ionic、Cordova 必要的软件，然后配置 Android、IOS 两种手机系统相关的打包环境。在 Windows 环境下可以进行 Android 系统的开发。在 Apple OS X 环境下可以进行 Android、IOS 等平台的开发。

1　Ionic 基本环境安装

（1）安装 Node.js

第一步：下载 Node.js。Node.js 最新版本的下载地址为：https://nodejs.org/en/。

第二步：查看版本号。打开 Node 命令窗口（Node.js command prompt），输入 node -v、npm -v 命令查看版本号，效果如图 1.5 和图 1.6 所示。

图 1.5　node -v　　　　　　　　　图 1.6　npm -v

（2）安装 Ionic

第一步：输入 npm install -g ionic 命令（该命令默认安装当下最新的 Ionic 版本）安装 Ionic。效果如图 1.7 所示。

图 1.7　安装 Ionic

提示：若需要安装指定版本的 Ionic 可使用 npm install -g ionic@3 命令（3 为 Ionic 的版本，可改为需要的其他版本）。

Ionic 安装完成后界面如图 1.8 所示。

图 1.8　Ionic 安装完成后效果图

第二步：输入 ionic -version 命令查看版本号，效果如图 1.9 所示。

```
C:\Users\Administrator.XTM-01706161027>ionic -version
3.7.0
```

图 1.9　查看版本号

（3）安装 Cordova

第一步：运行 npm install -g cordova 下载并安装 Cordova。效果如图 1.10 所示。

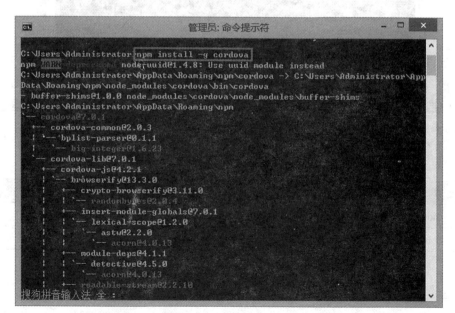

图 1.10　安装 Cordova

第二步：通过 cordova -v 查看是否安装成功，如果安装成功，则出现 Cordova 版本号，效果如图 1.11 所示。

图 1.11　查看 Cordova 版本号

（4）创建简单的 Ionic 项目

Ionic 基本环境配置完成，在命令窗口输入 ionic start MyIonicProject tabs 创建一个名 MyIonicProject 的项目。其中 start：开始创建一个新的应用程序；tabs：项目的模板样式。效果如图 1.12 所示。

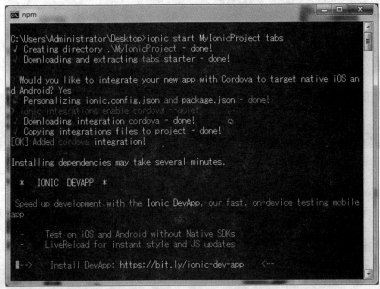

图 1.12　创建项目

除了使用 tabs 模板之外，Ionic 项目还可以使用其他的模板，一般不选择模板类型则默认使用选项卡模板（tabs）。模板类型如表 1.2 所示。

表 1.2　模板类型

类　　型	含　　义
sidemenu	侧面有可滑动菜单的布局
blank	一个单独的启动器
super	启动项目超过 14 个使用的页面设计
tutorial	一个引导启动项目

通过 cd MyIonicProject 命令进入创建的项目文件夹。使用 ionic serve 命令可以启动该项目，项目启动成功效果如图 1.13 所示，在浏览器上可以看到 Ionic 运行效果如图 1.14 所示。

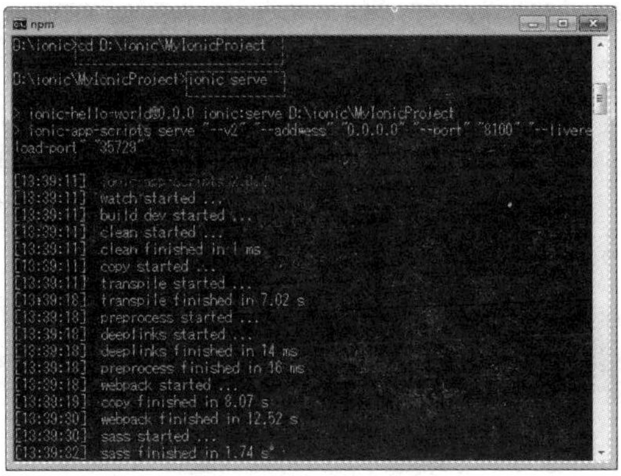

图 1.13 项目启动中

图 1.14 显示效果

2 在 Windows 下打包 APK

Ionic 开发 Android 手机系统的 APP 时,需要安装 JDK(JDK 的安装版本必须在 1.8 以上)和 SDK 进行项目打包。

(1)安装 JDK

(2)安装 SDK

SDK 又称作软件开发工具包,是被软件工程师用于为特定的软件包、软件框架、硬件平台、操作系统等建立应用软件的开发工具的集合。

第一步:下载 SDK。官网下载地址为(https://android-sdk.en.softonic.com/)。如图 1.15 所示。

图 1.15　SDK 官网

第二步：解压压缩包。

首先将 Android SDK 开发包解压，解压后，会得到以下几个重要的文件，但是在这里只选择"SDK Manager.exe"（负责下载和更新 SDK 包），如图 1.16 所示。

图 1.16　解压目录

第三步：选择版本安装。

自动检测是否有更新的 SDK 数据包可供下载，然后选择所需 Android 版本，点击"Install packages"安装。如图 1.17 所示。

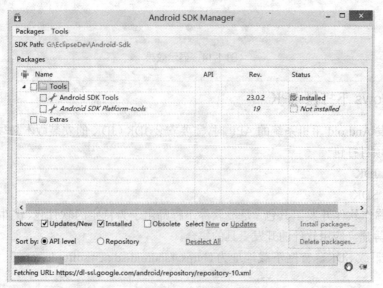

图 1.17　选择安装

第四步：配置环境变量。

将 SDK tools 目录的完整路径设置到系统变量中。新建变量名为"SDK_HOME"，在"变量值"文本框输入 Android SDK 解压路径，如图 1.18 所示。

图 1.18　设置系统变量

找到"Path"变量名，点击编辑，在"变量值"文本框中添加"%SDK_HOME%\tools"；如图 1.19 所示。

图 1.19　设置系统变量

第五步：查看安装版本。

设置完成后，检查 Android SDK 是否已经安装成功。在命令窗口输入"Android -h"，如果显示 Android SDK 的信息则证明安装成功，如图 1.20 所示。（注意：Android 和 -h 之间有空格。）

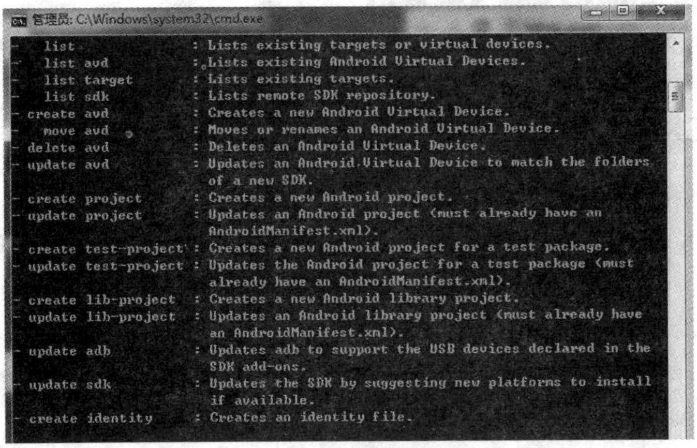

图 1.20　查看版本

在项目根目录下执行 ionic platform add android 命令添加一个指定的平台。该目录下会自动生成 platform 文件夹，里面包含 Android 文件夹。效果如图 1.21 所示。

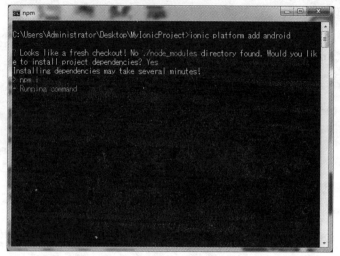

图 1.21　生成 platform 文件夹

在项目根目录下执行 ionic build android 命令，会在 xxx\platforms\android\build\outputs\apk 下生成 android-debug.apk，生成的 APK 用于调试。效果如图 1.22 和图 1.23 所示。

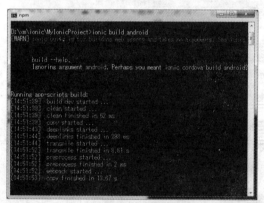

图 1.22　打包命令

图 1.23　生成 APK

3　在 Apple OSX 下打包 APK

构建 OSX 应用需要使用苹果提供的工具，这些工具只能在 OSX 操作系统的 Mac 设备上进行安装。可以使用 Xcode（Xcode 6.0 是最低版本）或者其他 IDE（例如 JetBrain 的 AppCode）来测试所有的 Cordova 特性。

（1）安装 Xcode

下载 Xcode 有两种方式：

● 在网页版 APP Store 或者 APP Store 应用中搜索"Xcode"。

● 从 Apple Developer Downloads 下载，需要注册 Apple 开发者。

Xcode 安装完之后，需要启用一些命令行工具来使 Cordova 正常运行。打开"Xcode"菜单，选择"Preferences"→"Downloads"→"Components"，然后点击"Command Line Tools"列表之后的"Install"按钮。

（2）运行 APP

使用 cordova run 命令运行 APP。使用"cordova run"→"help"命令来查看更多构建和运行选项。效果如图 1.24 所示。

图 1.24　运行 APP

（3）在 SDK 中打开项目

可以使用 Xcode 打开 OSX 平台。双击打开 hello/platforms/osx/HelloWorld.xcodeproj 文件。效果如图 1.25 所示。

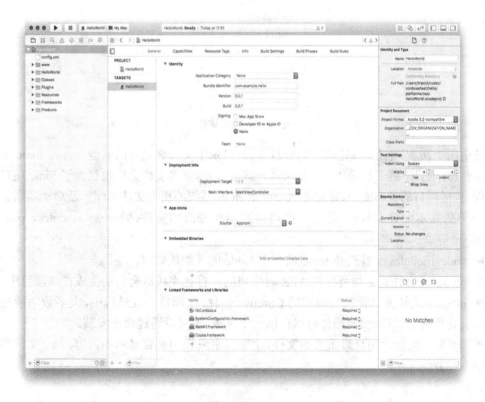

图 1.25　运行 APP

技能点 3 Ionic 项目结构

1 Ionic 项目结构

通过命令创建 Ionic 项目的效果如图 1.26 所示，那它们的结构又是怎样的呢？项目创建完成后的文件结构如图 1.26 所示。

图 1.26 项目结构

其中图 1.26 是项目下载完成后的根目录，根目录中文件夹之外的其他文件是 Ionic 项目的配置文件，在项目下载后自动生成。hooks、resource、www 等文件夹中也包含了 Ionic 项目的配置文件，包括项目的说明文档、项目的 Android、IOS 系统的相关文件和配置及启动动画。除了项目中一些基本配置文件外，还有一些可以利用和修改的文件，如 node_modules、platform、plugins、src 等。

- node_modules 文件夹是 Ionic 项目中 Node 的各类依赖包。
- platform 文件夹是项目打包 APK 时生成的。打包完成后会自动生成在该项目中。
- plugins 文件夹包含项目所需的 Cordova 插件，在项目开发过程中，大多数功能在实现时需要安装相应的插件，安装插件可以在网站上下载，也可以通过命令安装。
- src 文件夹包含项目最重要的文件，主要用来编辑和配置显示的内容。src 文件夹的结构如图 1.27 所示。

图 1.27 src 文件夹结构

- app：路由配置的文件夹；
- index.html：Ionic 应用的主入口文件；
- pages：项目中的所有界面的文件夹。

app 文件夹中包括 app.component.ts、app.module.ts、app.scss、main.ts、app.html。

（1）app.component.ts 是应用的入口文件，代码如下所示。

```typescript
import { Component } from '@angular/core';
import { Platform } from 'ionic-angular';
import { StatusBar } from '@ionic-native/status-bar';
import { SplashScreen } from '@ionic-native/splash-screen';
import { TabsPage } from '../pages/tabs/tabs';
@Component({
  templateUrl: 'app.html'
})
export class MyApp {
  rootPage:any = TabsPage;
  constructor(platform: Platform, statusBar: StatusBar, splashScreen: SplashScreen) {
    platform.ready().then(() => {
      statusBar.styleDefault();
      splashScreen.hide();
    });
  }
}
```

（2）app.module.ts 是定义应用程序的根模块，代码如下所示。

```typescript
import { NgModule, ErrorHandler } from '@angular/core';
import { BrowserModule } from '@angular/platform-browser';
import { IonicAPP, IonicModule, IonicErrorHandler } from 'ionic-angular';
```

```typescript
import { MyAPP } from './APP.component';
import { AboutPage } from '../pages/about/about';
import { ContactPage } from '../pages/contact/contact';
import { HomePage } from '../pages/home/home';
import { TabsPage } from '../pages/tabs/tabs';

import { StatusBar } from '@ionic-native/status-bar';
import { SplashScreen } from '@ionic-native/splash-screen';
@NgModule({
  declarations: [
    MyApp,
    AboutPage,
    ContactPage,
    HomePage,
    TabsPage
  ],
  imports: [
    BrowserModule,
    IonicModule.forRoot(MyAPP)
  ],
  bootstrap: [IonicAPP],
  entryComponents: [
    MyApp,
    AboutPage,
    ContactPage,
    HomePage,
    TabsPage
  ],
  providers: [
    StatusBar,
    SplashScreen,
    {provide: ErrorHandler, useClass: IonicErrorHandler}
  ]
})
export class APPModule {}
```

（3）app.html 是应用的主模版，在这个模板中，ion-nav 组件作为主要内容区域。[root] 是 Ionic 自带的 API（预先定义的函数）。

```
<ion-nav [root]="rootPage"></ion-nav>
```

pages 文件夹中的每个界面都是一个文件夹，文件夹中包括三个文件：.html 文件、.scss 文件、.ts 文件。

➢ .html 文件：用来编写项目界面的内容代码，代码如下所示。

```
<ion-header>
 <ion-navbar>
  <ion-title>Home</ion-title>
 </ion-navbar></ion-header>
<ion-content padding class="home">
 <h2>Ionic 2!</h2>
</ion-content>
```

➢ .scss 文件：主要用来编写界面的样式代码，代码如下所示。

```
.home {
  &.xxx{
  }
  .xxx{
  }
}
```

➢ .ts 文件：主要用来编写项目所需的方法，实现需要的功能，代码如下所示。

```
import {Component} from '@angular/core';import {NavController} from 'ionic-angular';
@Component({
 templateUrl: 'build/pages/home/home.html'
})
export class HomePage {
 constructor(private navCtrl: NavController) {
 }
}
```

2　Ionic 案例

页面跳转是 Ionic 项目中的常用功能。页面跳转需要在两个界面中实现，在主页面创建一个按钮，点击按钮，跳转到另一个页面。创建带有跳转功能的实例过程如下。

第一步：创建页面。

在项目文件夹下，使用"shift+ 鼠标右键"弹出快捷方式，选择"在此处打开命令窗口"，输入 ionic g page NewPage、ionic g page Hello（NewPage、Hello 为页面名字）命令创建界面。

第二步：页面配置。

在 app.module.ts 中配置新建页面地址，代码如下所示。

```
import { Hello } from '../pages/hello/hello';
import { NewPagePage } from '../pages/new-page/new-page';
@NgModule({ declarations: [
  MyApp,
  Hello,
  NewPagePage
], imports: [
  IonicModule.forRoot(MyApp)
],bootstrap: [IonicApp], entryComponents: [
  MyApp,
  Hello,
  NewPagePage
],providers: []
})
```

页面创建成功后，需要实现从主页面（hello.html）跳转到其他页面（new-page.html）的功能。在跳转过程中，也可以在页面之间传递数据。下面分基本跳转和带参数跳转讲解，实现过程如下：

（1）基本跳转

第一步：在 hello.html 文件中添加 button 标签并给 button 标签添加点击事件。代码如下所示。

```
<button class="hello-ionic-btn" (click)="testNewPage()">
    <ion-icon name="menu"></ion-icon>
</button>
```

第二步：在 hello.ts 文件中添加页面跳转方法，代码如下所示。

```
import { NewPagePage } from '../new-page/new-page';
constructor(public navCtrl: NavController) {
}
testNewPage(){
  console.log(' 点击跳转 ');
  this.navCtrl.push(NewPagePage);
}
```

（2）带参数跳转

第一步：在 hello.html 文件中添加 button 标签并添加跳转相关参数，具体代码如下所示。

```
<button ion-item [navPush]="nxPage" [navParams]="params">
点击跳转
</button>
```

第二步：在 hello.ts 文件中配置页面地址及参数，具体代码如下所示。

```
import { NewPagePage } from '../new-page/new-page';
nxPage: any = NewPagePage;
params: any = {id: 42};
```

第三步：在新界面取值，可通过两种方式取值，分别是通过 navCtrl.push 取值和 [navParams] 取值。

➢ 通过 navCtrl.push 传过来的取值，具体代码如下所示。

```
import { NavController,NavParams } from 'ionic-angular';
export class NewPagePage {
  id :any;
  constructor(public navCtrl: NavController,public navParams: NavParams) {
  console.log(this.navParams.get('id'));
  }
}
```

➢ 通过页面 [navParams] 传过来的取值，具体代码如下所示。

```
export class NewPagePage {
  data:any;
  constructor(public navCtrl: NavController,public navParams: NavParams) {
    this.data = navParams.data
    console.log(this.data);
  }
  ionViewDidLoad(navParams: NavParams) {
    console.log('Hello NewPagePage Page');
    console.log(this.data);
  }
}
```

提示 想了解或学习更多的关于 Ionic3 相对于 Ionic2 的优势，扫描图中二维码，获得更多信息。

通过下面九个步骤的操作，实现图 1.2 和图 1.4 所示的"Dancer 时代"登录界面及所对应的功能。

第一步：打开命令窗口，进入想要创建项目的目录下，运行 ionic start myApp blank 创建一个空白的项目，如图 1.28 所示。

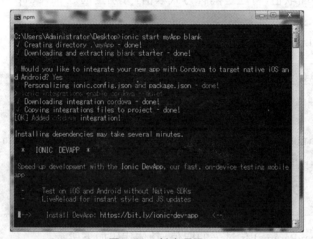

图 1.28 创建项目

第二步：进入项目所在路径输入命令创建 CORE0101 页面。如图 1.29 所示。

图 1.29 创建页面

第三步：在 app.module.ts 中进行配置，引入 CORE0101 页面，并将其导入 @NgModule 模块中，代码如 CORE0102 所示。

代码 CORE0102　　app.module.ts

```typescript
import { NgModule, ErrorHandler } from '@angular/core';
import { BrowserModule } from '@angular/platform-browser';
import { IonicApp, IonicModule, IonicErrorHandler } from 'ionic-angular';
import { MyApp } from './app.component';
import { HomePage } from '../pages/home/home';
import { Core0101Page } from '../pages/CORE0101/CORE0101';
import { StatusBar } from '@ionic-native/status-bar';
import { SplashScreen } from '@ionic-native/splash-screen';
// 需要配置的 ts 路径
@NgModule({
  declarations: [
    MyApp,
    HomePage,
    Core0101Page // 必须引入的参数名
  ],
  imports: [
    BrowserModule,
    IonicModule.forRoot(MyApp)
  ],
  bootstrap: [IonicApp],
  entryComponents: [
    MyApp,
    HomePage,
    Core0101Page // 必须引入的参数名
  ],
  providers: [
    StatusBar,
    SplashScreen,
    {provide: ErrorHandler, useClass: IonicErrorHandler}
  ]
})
export class AppModule {}
```

第四步：在 app.component.ts 中进行加载的配置，需要将 CORE0101 页面引入并设置它为第一个加载的页面，代码如 CORE0103 所示。

代码 CORE0103　app.component.ts

```
import { Component } from '@angular/core';
import { Platform } from 'ionic-angular';
import { StatusBar } from '@ionic-native/status-bar';
import { SplashScreen } from '@ionic-native/splash-screen';
// 需要配置的 ts 路径
import { Core0101Page } from '../pages/CORE0101/CORE0101';
@Component({
  templateUrl: 'app.html'
})
export class MyApp {
  rootPage:any = Core0101Page;// 第一个被加载页面的配置参数
  constructor(platform: Platform, statusBar: StatusBar, splashScreen: SplashScreen) {
    platform.ready().then(() => {
      statusBar.styleDefault();
      splashScreen.hide();
    });
  }
}
```

第五步：在 CORE0101.ts 中更改 class 类的名称（跟 app.module.ts 中名称统一），代码如 CORE0104 所示。

代码 CORE0104　CORE0101.ts

```
import { Component } from '@angular/core';
import { NavController, NavParams } from 'ionic-angular';
@Component({
  selector: 'page-core0101',
  templateUrl: 'CORE0101.html',
})
export class Core0101Page {
  //Core0101Page 必须和 app.module.ts 中配置的 html 路径名称相同
  constructor(public navCtrl: NavController, public navParams: NavParams) {
  }
}
```

第六步：登录首界面的制作。

登录首界面是由背景图和按钮组成，使用 <button> 标签在背景图的上方添加按钮。代码 CORE0105 如下。设置样式前效果如图 1.30 所示。

项目一 "Dancer 时代"登录模块的实现

代码 CORE0105　登录首界面（CORE0101.html）

```
<ion-content class="page-core0101">
  <div class=" div">
    <img src="assets/img/1.jpg ">
  </div>
  <!--ion-button block 为 ionic 预定义样式；-->
  <button ion-button block (click)="register($event)" class=" regis">
    注册
  </button>
  <button ion-button block (click)="login($event)" class="log">
    登录
  </button>
</ion-content>
```

图 1.30　登录首界面设置样式前

设置登录首界面样式，需要为背景图片设置宽高来改变图片的大小，并调整按钮的位置。部分代码 CORE0106 如下。设置样式后效果如图 1.31 所示。

代码 CORE0106　登录首界面样式（CORE0101.scss）

```
*{
  margin: 0;
  padding: 0;
}
// 去掉所有的边框样式
```

```css
.div{
  height=100%;
  width=100%;
}
//div 的高度、宽度设置
.div img{
  height=100%;
  width=100%;
}
// 注册按钮的样式
.regis{
  position: absolute;
  bottom: 0px;
  margin:0;
  left: 0%;
  width: 50%;
  border-radius: 3px;
  background: #cccccc;
  z-index: 10;
}
// 登录按钮的样式
.log{
  position: absolute;
  width: 50%;
  margin:0;
  right:0;
  bottom: 0px;
  border-radius: 3px;
  background: red;
  z-index: 10;
}
```

第七步：在命令窗口运行 ionic g page login 创建登录页面，在 app.module.ts 中配置，代码如 CORE0107 所示。

图 1.31 登录首界面设置样式后

代码 CORE0107　app.module.ts

```
import { NgModule, ErrorHandler } from '@angular/core';
import { BrowserModule } from '@angular/platform-browser';
import { IonicApp, IonicModule, IonicErrorHandler } from 'ionic-angular';
import { MyApp } from './app.component';
import { HomePage } from '../pages/home/home';
import { Core0101Page } from '../pages/CORE0101/CORE0101';
import { LoginPage } from '../pages/login/login';
import { StatusBar } from '@ionic-native/status-bar';
import { SplashScreen } from '@ionic-native/splash-screen';
// 需要配置的 ts 路径
@NgModule({
  declarations: [
    MyApp,
    HomePage,
    LoginPage,
    Core0101Page // 必须引入的参数名
  ],
  imports: [
    BrowserModule,
    IonicModule.forRoot(MyApp,{
```

```
      tabsHideOnSubPages: 'true'        // 隐藏全部子页面 tabs
    })
  ],
  bootstrap: [IonicApp],
  entryComponents: [
    MyApp,
    HomePage,
    LoginPage,
    Core0101Page // 必须引入的参数名
  ],
  providers: [
    StatusBar,
    SplashScreen,
    {provide: ErrorHandler, useClass: IonicErrorHandler}
  ]
})
export class AppModule {}
```

在 login.ts 中配置,代码如 CORE0108 所示。

代码 CORE0108　login.ts

```
import { Component } from '@angular/core';
import { NavController } from 'ionic-angular';
@Component({
  selector: 'page-list',
  templateUrl: 'login.html'
})
export class LoginPage {
  constructor(public navCtrl:NavController) {}
}
```

第八步:登录首界面设置页面跳转效果,当点击登录按钮时,跳转到登录界面。部分代码如 CORE0109 所示。

代码 CORE0109　设置页面跳转(CORE0101.ts)

```
import { Component } from '@angular/core';
import { NavController, NavParams } from 'ionic-angular';
import { LoginPage } from '../login/login';
```

```
// 导入登录界面
@Component({
  selector: 'page-core0101',
  templateUrl: 'CORE0101.html',
})
export class Core0101Page {
  //Core0101Page 必须和 app.module.ts 中配置的 html 路径名称相同
  constructor(public navCtrl: NavController, public navParams: NavParams) {
  }
  login(){
    this.navCtrl.push(LoginPage);
    // 跳转到登录页面
  }
}
```

第九步:登录界面的制作。

登录界面是由用户的头像和 input 输入框组成,代码 CORE0110 如下。设置样式前效果如图 1.32 所示。

代码 CORE0110　登录界面(login.html)

```html
<ion-header>
  <ion-navbar>
    <ion-title> 登录 </ion-title>
  </ion-navbar>
</ion-header>
<ion-content has-bouncing="true" overflow-scroll="false">
  <div>
    <img class="loginimg" src="assets/img/ionic.png">
    <!-- 顶部图片 -->
  </div>
  <ion-list>
    <ion-item>
      <ion-label stacked>Username:</ion-label>
      <ion-input type="text" placeholder=" 请输入用户名 " [(ngModel)]="username" clearInput=true></ion-input>
      <!--[(ngModel)] 进行数据的双向绑定 -->
    </ion-item>
```

```
            <ion-item>
               <ion-label stacked>Password:</ion-label>
               <ion-input type="password" placeholder=" 请输入密码 " [(ngModel)]="password" clearInput=true></ion-input>
            </ion-item>
         <!--</form>-->
         </ion-list>
         <button ion-button full (click)="login()">
            登录
         </button>
      </ion-content>
```

图 1.32　登录界面设置样式前

设置登录界面样式，对头像进行圆角设置并将其居中。部分代码 CORE0111 如下。设置样式后效果如图 1.33 所示。

代码 CORE0111　登录界面样式（login.scss）

// 设置图片的圆角并居中显示
.loginimg{
　width: 150px;
height:150px;
　margin-top: 20px;

```
  border-radius: 50%;
  margin-left: -75px;
  position: relative;
  left: 50%;
}
// 设置 input 提示文字的大小
.logininput{
  padding:0;
  font-size: 12px;
}
```

图 1.33　登录界面设置样式后

当点击登录按钮后,获取输入框中数据并进行格式验证,以减少对后台的请求次数。部分代码 CORE0112 如下。

代码 CORE0112　登录界面交互(login.ts)

```
import { Component } from '@angular/core';
import { NavController } from 'ionic-angular';
import { TabsPage } from '../tabs/tabs';
@Component({
  selector: 'page-list',
  templateUrl: 'login.html'
})
```

```
export class LoginPage {
  username:string;
  // 定义 username 类型
  password:string;
  constructor(public navCtrl:NavController) {}
  login() {
    var user = this.username;
    // 给 user 变量赋值
    //this.username 为 input 中的 value 值
    var pw = this.password;
    // 给 pw 变量赋值
    //this.password 为 input 中的 value 值
    if (/^(13[0-9]|15[012356789]|17[03678]|18[0-9]|14[57])[0-9]{8}$/.test(user)) {
      // 正则表达式判断手机号
      if (/\w{6,16}/.test(pw)) {
        // 正则表达式判断密码格式
        if (user == "13752525756" && pw == "123456") {
          this.navCtrl.push(TabsPage);
          // 页面跳转
        } else {
          console.log(" 用户名或密码错误 ")
        }
      } else {
        console.log(" 密码格式为 6-16 为数字字母下划线的组合 ")
      }
    } else {
      console.log(" 请正确填写手机号 ")
    }
  }
}
```

至此，"Dancer 时代"登录模块制作完成。

本项目通过"Dancer 时代"登录模块的学习，能够对跨平台开发技术的介绍及特性有所认识，对 Ionic 环境搭建及 APK 打包具有初步了解，同时能够掌握 Ionic 的项目结构及具有使用 Ionic 创建界面的本领。

mobile	移动	advanced	先进的
framework	框架	buttons	按钮
beta	公开测试	Issue	问题
blank	空白的	resource	资源

一、选择题

1. 以下哪个不是跨平台开发框架（　　）。
A.Mobile Angular UI　B.Ionic　　　　C.Tenga　　　　　　D.Sencha Touch
2. 以下哪个不是 Ionic 的优点（　　）。
A. 漂亮的界面，追求性能，专注原生，免费开源
B. 数据集成
C. 数据双向绑定
D. 基于 Cordova
3. 本书中 Node.js 使用版本为（　　）。
A.7.x　　　　　　B.8.x　　　　　　C.6.x　　　　　　　D.4.x
4. 使用命令行创建 APP 时，默认会使用（　　）模板。
A.tutorial　　　　B.tabs　　　　　　C.blank　　　　　　D.side
5. Android 系统的移动应用开发需要在 Window 环境下进行，开发前需要安装 Node.js、
（　　）、CLI、Cordova、JDK、SDK 软件并配置相关的环境。
A.Ionic　　　　　B.install　　　　　C.npm　　　　　　D.cnpm

二、填空题

1. Mobile Angular UI 是使用 _____ 和 _____ 的响应式移动开发 HTML5 框架。
2. Ionic 是一个强大的 _____ 应用程序开发框架。
3. Ionic3 项目在编辑过程中在 _____ 文件夹下面进行。
4. Ionic3 有自己的类（Class）、_____ 和自己的样式文件（在这里我们提倡使用 scss）。
5. Ionic1.xh 和 Ionic3 之间的区别包括 _____、命令行工具、路由导航、_____。

三、上机题

使用 Ionic 知识实现下列要求的效果。要求：
1. 使用 Ionic3 创建一个空项目。
2. 创建两个页面并实现页面之间的跳转。

项目二 "Dancer 时代"首页模块的实现

通过"Dancer 时代"首页模块的实现,了解首页的设计理念及功能实现,学习 Ionic 相关组件及如何应用列表实现布局,掌握 Ionic 使用轮播组件实现首页模块的轮播,具有使用 Ionic 组件实现首页布局的能力,在任务实现过程中:

- 了解 Ionic 的相关组件。
- 掌握 Ionic 列表。
- 掌握 Ionic 使用轮播图的步骤。
- 具有使用 Ionic 组件实现首页布局的能力。

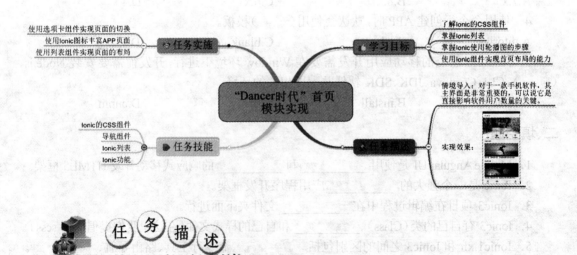

【情境导入】

对于一款手机软件,其首页是非常重要的,可以说它是直接影响软件用户数量的关键。所以界面设计人员 Ellison 在首页的设计和美化上花了很大的心思,他在设计首页时不仅考虑到软件信息的宣传还考虑了用户使用软件的便利性,所以 Ellison 在界面添加了手机软件中常用的轮播图元素。首页中还有视频和音乐的推荐功能,点击跳转对应界面可进行视频和音频的播放。本项目主要通过实现"Dancer 时代"的首页了解 Ionic 的轮播图和导航组件。

【功能描述】

本项目将实现"Dancer 时代"首页模块。
- 使用选项卡组件实现页面的切换。
- 使用 Ionic 图标丰富 APP 页面。
- 使用列表组件实现页面的布局。

【基本框架】

基本框架如图 2.1 所示。通过本项目的学习,能将框架图 2.1 转换成效果图 2.2。

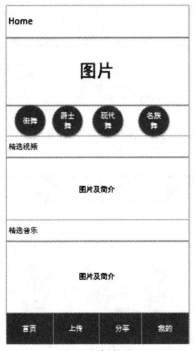

图 2.1　框架图

图 2.2　效果图

技能点 1　Ionic 的 CSS 组件

1　Ionic 字体图标

Ionic 使用 ionicons 图标样式库。ionicons 采用了 TrueType 字体实现图标样式,有超过

700 个图标可供选择。使用字体图标的代码结构如下:

```
<ion-icon name=" 图片属性名 "></ion-icon>
```

Ionic 字体图标具有两个状态:激活和未激活。可以通过设置属性 isActive 为 true 或 false 更改图标的状态。如果没有指定值,图标默认显示为激活状态,代码如下所示:

```
<!-- 激活状态 -->
<ion-icon name="heart"></ion-icon>
<!-- 未激活状态 -->
<ion-icon name="heart" isActive="false"></ion-icon>
```

提示:如果想指定平台的图标,需要使用 ios 或 md 属性,并提供平台特定的图标名称。

```
<!-- 使用 ios 和 android 系统图标 -->
<ion-icon ios="logo-apple" md="logo-android"></ion-icon>
```

当固定图标不能满足项目要求时,需要动态设置图标的样式,此时需要使用变量设置字体图标,对应的 HTML 代码如下:

```
<ion-icon [name]="myIcon"></ion-icon>
```

对应的 ts 代码如下:

```
export class MyFirstPage {
  myIcon: string = "home";
}
```

提示:可以通过 http://ionicframework.com/docs/Ionicons/ 查看所有的字体图标集。

2 按钮

按钮是界面设计中不可缺少的元素之一,不同风格的 APP 需要的按钮样式也不相同。与 Bootstrap 前端库类似,Ionic 提供大量的按钮样式,能够满足大部分的界面需求。默认情况下,按钮样式为 display: inline-block。Ionic 按钮属性如表 2.1 所示。

表 2.1 Ionic 按钮属性

属 性	说 明
outline	使用 outline 属性设置轮廓线样式
clear	使用 clear 属性清除按钮属性
round	使用 round 属性设置按钮圆形边框
block	使用 block 属性设置按钮占据容器宽度的 100%

续表

属　　性	说　　明
full	使用 full 属性设置按钮占据容器宽度的 100%

使用按钮效果如图 2.3 所示。

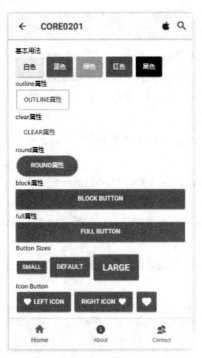

图 2.3　按钮效果示例

为了实现图 2.3 的效果，新建 CORE0201.html，代码如下所示。

代码 CORE0201　按钮

```
<ion-header>
  <ion-navbar>
    <ion-title>CORE0201</ion-title>
    <ion-buttons start>
      <button ion-button icon-only>
        <ion-icon name="logo-apple"></ion-icon>
      </button>
    </ion-buttons>
    <ion-buttons end>
    <button ion-button icon-only>
      <ion-icon name="search"></ion-icon>
    </button>
```

```html
    </ion-buttons>
  </ion-navbar>
</ion-header>
<ion-content padding>
  <div> 基本用法 </div>
    <button ion-button color="light"> 白色 </button>
    <button ion-button> 蓝色 </button>
    <button ion-button color="secondary"> 绿色 </button>
    <button ion-button color="danger"> 红色 </button>
    <button ion-button color="dark"> 黑色 </button>
  <div>outline 属性 </div>
    <button ion-button outline>OUTLINE 属性 </button>
  <div>clear 属性 </div>
    <button ion-button clear>CLEAR 属性 </button>
  <div>round 属性 </div>
    <button ion-button round>ROUND 属性 </button>
  <div>block 属性 </div>
    <button ion-button block>Block Button</button>
  <div>full 属性 </div>
    <button ion-button full>Full Button</button>
  <div>Button Sizes</div>
  <!-- 比默认按钮小 -->
    <button ion-button small>SMALL</button>
  <!-- 默认按钮 -->
  <button ion-button>DEFAULT</button>
  <!-- 比默认按钮大 -->
    <button ion-button large>LARGE</button>
  <div>Icon Button</div>
  <!-- 字体图标在左边 -->
    <button ion-button icon-left>
<ion-icon name="heart"></ion-icon>
LEFT ICON
    </button>
  <!-- 字体图标在右边 -->
    <button ion-button icon-right>
RIGHT ICON
    <ion-icon name="heart"></ion-icon>
    </button>
```

```
    <button ion-button icon-only>
        <ion-icon name="heart"></ion-icon>
    </button>
</ion-content>
```

3 Ionic 表单

在手机 APP 设计中，输入框、单选框、复选框、Toggle 等来自表单控件，大多数情况被应用于登录、注册、设置等界面。

（1）输入框

输入框是一个常用于注册、登录界面的表单提交控件。在 Ionic 中，通过使用 ion-input 标签可以收集和处理用户输入的信息。Ionic 为界面设计提供了多种输入框样式，Ionic 输入框属性如表 2.2 所示。

表 2.2　Ionic 输入框属性

属　　性	说　　明
fixed labels	使用 fixed 属性在 input 元素左侧放一个 label，input 将会对齐到相同的位置
floating labels	使用 floating 属性，当选择输入的时候 lable 会浮动到上方
inline labels	没有任何属性的 ion-label 是一个 inline label，输入的时候不会消失
inset labels	每个输入项默认占据父级元素 100% 宽度
placeholder labels	为 input 元素添加 placeholder 属性，当输入文本的时候，占位文本会消失
stacked labels	stacked label 通常会显示在 input 上方

使用输入框效果如图 2.4 所示。

图 2.4　输入框

为了实现图 2.4 的效果，新建 CORE0202.html，代码如下所示。

```
代码 CORE0202    输入框
<ion-header>
  <ion-navbar>
    <ion-title>CORE0202</ion-title>
  </ion-navbar>
</ion-header>
<ion-content padding>
  <ion-list inset>
    <ion-item>
      <ion-label fixed >用户</ion-label>
      <ion-input type="text" value=""></ion-input>
    </ion-item>
      <ion-item>
      <ion-label floating> 昵称 </ion-label>
      <ion-input type="password"></ion-input>
    </ion-item>
      <ion-item>
      <ion-label> 密码 </ion-label>
      <ion-input type="text"></ion-input>
    </ion-item>
      <ion-item>
      <ion-input type="text" placeholder=" 邮箱地址 "></ion-input>
    </ion-item>
      <ion-item>
      <ion-label stacked> 电话号码 </ion-label>
      <ion-input type="text"></ion-input>
    </ion-item>
  </ion-list>
</ion-content>
```

（2）单选框

Ionic 提供 item-radio 组件来表示单选框。该 item-radio 组件与 HTML 中的 radio 类似，每个 item-radio 都具有相同的 name 属性，一次只能选择一个选项。通常情况下 item-radio 使用 checked 属性设置默认选中值，使用 disabled 属性设置禁止用户修改。使用单选框效果如图 2.5 所示。

项目二 "Dancer 时代"首页模块的实现

图 2.5 单选框

为了实现图 2.5 的效果,新建 CORE0203.html,代码如下所示。

代码 CORE0203　单选框

```
<ion-header>
 <ion-navbar>
  <ion-title>CORE0203</ion-title>
   </ion-navbar>
</ion-header>
<ion-content padding>
 <ion-list radio-group>
  <ion-list-header>
   运动
  </ion-list-header>
  <!-- 默认值 -->
  <ion-item>
    <ion-label> 篮球 </ion-label>
    <ion-radio checked="true" value=" 篮球 "></ion-radio>
  </ion-item>
  <ion-item>
    <ion-label> 足球 </ion-label>
    <ion-radio value=" 足球 "></ion-radio>
```

```
        </ion-item>
        <!-- 禁止使用 -->
        <ion-item>
            <ion-label> 羽毛球 </ion-label>
            <ion-radio value=" 羽毛球 " disabled="true"></ion-radio>
        </ion-item>
    </ion-list>
</ion-content>
```

(3)复选框

Ionic 的复选框(checkbox)组件与 HTML 的 checkbox 用法类似,可以通过 checked 设置默认值,使用 disabled 属性设置禁止用户修改,checkbox-assertive 指定复选框的颜色和样式。使用复选框效果如图 2.6 所示。

图 2.6 复选框

为了实现图 2.6 的效果,新建 CORE0204.html,代码如下所示。

代码 CORE0204　复选框

```
<ion-header>
    <ion-navbar>
        <ion-title>CORE0204</ion-title>
    </ion-navbar>
</ion-header>
```

```
        <ion-content padding>
      <ion-list>
       <ion-list-header>
        运动
       </ion-list-header>
       <ion-item>
         <ion-label> 篮球 </ion-label>
         <ion-checkbox checked="true"></ion-checkbox>
       </ion-item>
       <ion-item>
         <ion-label> 足球 </ion-label>
         <ion-checkbox color="dark" checked="true"></ion-checkbox>
       </ion-item>
         <ion-item>
         <ion-label> 羽毛球 </ion-label>
         <ion-checkbox value="cherry" disabled="true"></ion-checkbox>
       </ion-item>
       <ion-item>
         <ion-label> 乒乓球 </ion-label>
         <ion-checkbox color="secondary"></ion-checkbox>
       </ion-item>
       <ion-item>
         <ion-label> 棒球 </ion-label>
         <ion-checkbox color="danger" checked="true"></ion-checkbox>
       </ion-item>
      </ion-list>
     </ion-content>
```

（4）Toggle

Toggle 与复选框类似。但 Toggle 更容易在移动设备上使用，经常用来打开或关闭某个选项。常用的属性有 value、disabled、checked 等。使用 Toggle 效果如图 2.7 所示。

图 2.7 Toggle

为了实现图 2.7 效果,新建 CORE0205.html,代码如下所示。

```
代码 CORE0205    Toggle
<ion-header>
 <ion-navbar>
  <ion-title>CORE0205</ion-title>
 </ion-navbar>
</ion-header>
<ion-content padding>
 <ion-list>
  <ion-list-header>
    运动
  </ion-list-header>
  <ion-item>
    <ion-label> 篮球 </ion-label>
    <ion-toggle  checked="true"></ion-toggle>
  </ion-item>
  <ion-item>
    <ion-label> 足球 </ion-label>
    <ion-toggle color="energized"></ion-toggle>
```

```
        </ion-item>
        <ion-item>
          <ion-label> 羽毛球 </ion-label>
          <ion-toggle color="danger" checked="true"></ion-toggle>
        </ion-item>
        <ion-item>
          <ion-label> 乒乓球 </ion-label>
          <ion-toggle color="royal" checked="true"></ion-toggle>
        </ion-item>
        <ion-item>
          <ion-label> 游泳 </ion-label>
          <ion-toggle color="danger"></ion-toggle>
        </ion-item>
      </ion-list>
</ion-content>
```

提示 在网站开发的过程中，如果想要使表单中填入的信息更规范，扫描下方二维码，将会为你的网站添加更多的色彩。快来扫我吧！

技能点 2　导航组件

1　Ionic 选项卡

选项卡包括几个选项（选项的数量最多定义 5 个），用户可以点击选项来切换界面。选项卡通过 ion-tabs 标签定义，选项使用 ion-tab（ion-tab 是 ion-tabs 的子标签，每个选项卡的选项都包含一个相对应的界面）标签定义。Ionic 提供多种样式选项卡，样式属性如表 2.3 所示。

表 2.3　Ionic 选项卡样式属性

属　　性	说　　明	使用方法
tabsHighlight	设置 tab 标签是否高亮选中效果，默认值为 false	\<ion-tabs tabsHighlight="true"\>\</ion-tabs\>
tabsPlacement	设置选项卡出现的位置，属性值分别有：top、bottom，默认值为 bottom	\<ion-tabs tabsPlacement="top"\>\</ion-tabs\>

续表

属 性	说 明	使用方法
tabIcon	设置 tabIcon 属性可以在每个选项卡中添加一个图标	\<ion-tabs tabIcon="contact">\</ion-tabs>
tabsLayout	设置 tabsLayout 属性可以修改图标和文字的显示位置,如:icon-top 图标在上、icon-hide 隐藏图标等	\<ion-tabs tabsLayout="icon-top">\</ion-tabs>
selectedIndex	设置 selectedIndex 属性是默认选择的选项卡的索引值默认选中的 tabs(从 0 开始)	\<ion-tabs selectedIndex="2">\</ion-tabs>

使用选项卡效果如图 2.8 所示。

图 2.8　Ionic 选项卡

为了实现图 2.8 的效果,新建 tabs.html、home.html、about.html、contact.html,其中 tabs.html 代码如 CORE0206 所示。

代码 CORE0206　tabs.html

```
<ion-tabs>
    <!--root 属性是用来设置选项所对应的界面,title 属性用来设置选项卡显示的文字 -->
    <ion-tab [root]="tab1Root" tabTitle="Home" tabIcon="home"></ion-tab>
    <ion-tab [root]="tab2Root" tabTitle="About" tabIcon="information-circle"></ion-tab>
    <ion-tab [root]="tab3Root" tabTitle="Contact" tabIcon="contacts"></ion-tab>
</ion-tabs>
```

对 tabs.html 进行配置,代码如 CORE0207 所示。

代码 CORE0207 tabs.ts

```typescript
import { Component } from '@angular/core';
import { AboutPage } from '../about/about';
import { ContactPage } from '../contact/contact';
import { HomePage } from '../home/home';

@Component({
  templateUrl: 'tabs.html'
})
export class TabsPage {
// 三个选项对应切换的界面,与 root 属性的属性值相对应
  tab1Root = HomePage;
  tab2Root = AboutPage;
  tab3Root = ContactPage;
  constructor() {

  }
}
```

home.html 代码如 CORE0208 所示。

代码 CORE0208 home.html

```html
<ion-header>
  <ion-navbar>
    <ion-title>Home</ion-title>
  </ion-navbar>
</ion-header>
<ion-content padding>
  <h2>欢迎使用 Ionic</h2>
</ion-content>
```

界面的路由配置,代码如 CORE0209 所示。

代码 CORE0209 app.module.ts

```typescript
import { NgModule, ErrorHandler } from '@angular/core';
import { BrowserModule } from '@angular/platform-browser';
import { IonicApp, IonicModule, IonicErrorHandler } from 'Ionic-angular';
import { MyApp } from './app.component';
```

```
// 三个选项所对应的界面的路由配置
import { AboutPage } from '../pages/about/about';
import { ContactPage } from '../pages/contact/contact';
import { HomePage } from '../pages/home/home';
// 选项卡界面的路由配置
import { TabsPage } from '../pages/tabs/tabs';
import { StatusBar } from '@Ionic-native/status-bar';
import { SplashScreen } from '@Ionic-native/splash-screen';
@NgModule({
  declarations: [
    MyApp,
    AboutPage,
    ContactPage,
    HomePage,
    TabsPage
  ],
  imports: [
    BrowserModule,
    IonicModule.forRoot(MyApp)
  ],
  bootstrap: [IonicApp],
  entryComponents: [
    MyApp,
    AboutPage,
     ContactPage,
      HomePage,
    TabsPage
  ],
  providers: [
    StatusBar,
    SplashScreen,
    {provide: ErrorHandler, useClass: IonicErrorHandler}
  ]
})
export class AppModule {}
```

2 Ionic 侧栏菜单

Ionic 提供的侧栏菜单导航是通过点击带有 menuToggle 元素的标签实现的，默认情况下从左侧划入，可以通过设置 side="right" 实现从右侧滑入。侧栏菜单的作用是为界面节省空间，方便用户查询。侧边菜单属性如表 2.4 所示。

表 2.4　Ionic 侧栏菜单属性

属　　性	说　　明
menuToggle	从模板切换到打开或关闭
menuClose	从模板中关闭侧拉菜单
enabled	菜单是否可用
side	设置从左、从右滑入，默认从左滑入
type	菜单显示的类型，默认是根据模式的不同而不同

使用侧栏菜单效果如图 2.9 和图 2.10 所示。

图 2.9　侧栏菜单首页　　　　　　　　　图 2.10　侧栏菜单效果

为了实现图 2.9 和图 2.10 效果，新建界面，代码如 CORE0210 所示。

代码 CORE0210　侧栏菜单

```
<ion-header>
 <ion-navbar>
   <button ion-button menuToggle>
     <ion-icon name="menu"></ion-icon>
   </button>
```

```html
      <ion-title>CORE0206</ion-title>
    </ion-navbar>
  </ion-header>
  <ion-content padding>
    <h3>Welcome to your first Ionic app!</h3>
    <p>
      This starter project is our way of helping you get a functional app running in record time.
    </p>
    <p>
      Follow along on the tutorial section of the Ionic docs!
    </p>
    <p>
      <button ion-button color="primary" menuToggle>Toggle Menu</button>
    </p>
  </ion-content>
```

对应的侧栏菜单显示内容界面,代码如 CORE0211 所示。

代码 CORE0211 app.html

```html
<ion-menu [content]="content">
  <ion-header>
    <ion-toolbar>
      <ion-title>Pages</ion-title>
    </ion-toolbar>
  </ion-header>
  <ion-content>
    <ion-list>
      <button ion-item>
        我的文件
      </button>
      <button ion-item>
        我的钱包
      </button>
      <button ion-item>
        我的收藏
      </button>
      <button ion-item>
```

```
          我的会员
        </button>
      </ion-list>
    </ion-content>
  </ion-menu>
  <ion-nav [root]="rootPage" #content swipeBackEnabled="false"></ion-nav>
```

提示 你想知道如何使用导航组件进行页面之间的跳转吗？扫描下方二维码，你将会收获更多，心动不如行动，快来扫我吧！

技能点 3　Ionic 列表

1　基本列表

列表主要用于显示行信息，如联系人列表、播放列表等。列表不仅可以使用自身的属性设置样式和内容，还可以自定义样式或与其他组件（基本文本、按钮、切换开关、图标和缩略图等）配合使用。

默认情况下，基本列表之间有分隔线，若要隐藏列表项之间的分隔线，只需在 <ion-list> 标签内添加 no-lines 属性。使用基本列表效果如图 2.11 所示。

图 2.11　基本列表

为了实现图 2.11 效果，新建界面，代码如 CORE0212 所示。

代码 CORE0212　基本列表

```html
<ion-header>
  <ion-navbar>
    <ion-title> 列表 </ion-title>
   </ion-navbar>
</ion-header>
<ion-content>
  <ion-list>
    <button ion-item *ngFor="let item of items" (click)="itemSelected(item)">
      {{ item }}
    </button>
  </ion-list>
</ion-content>
```

对应 ts 文件，代码如 CORE 0213 所示。

代码 CORE0213　ts 文件

```typescript
import { Component } from '@angular/core';
@Component({
  templateUrl: 'hello-ionic.html'
})
export class HelloIonicPage {
  items = [
    ' 列表 1',
    ' 列表 2',
    ' 列表 3',
    ' 列表 4',
    ' 列表 5',
    ' 列表 6',
    ' 列表 7',
    ' 列表 8',
    ' 列表 9',
    ' 列表 10',
    ' 列表 11'
  ];
  itemSelected(item: string) {
```

```
      console.log("Selected Item", item);
   }
}
```

2 列表分隔符

列表分隔符可以用来将列表划分成多个列表组。好处是使浏览者更方便查询列表项,从而准确定位到所需信息。其实现方法是使用 <ion-item-group> 标签作为列表容器,通过 <ion-item-divider> 标签将列表划分为多个部分。使用列表分隔符效果如图 2.12 所示。

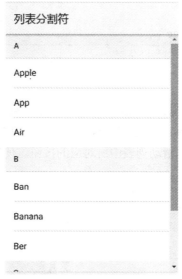

图 2.12 列表分隔符

为了实现图 2.12 效果,新建界面,代码如 CORE0214 所示。

```
代码 CORE0214   列表分隔符
<ion-header>
  <ion-navbar>
    <ion-title> 列表分割符 </ion-title>
  </ion-navbar>
</ion-header>
<ion-content>
  <ion-item-group>
    <ion-item-divider color="light">A</ion-item-divider>
    <ion-item>Apple</ion-item>
    <ion-item>App</ion-item>
    <ion-item>Air</ion-item>
```

```
    </ion-item-group>
    <!--省略部分代码 -->
</ion-content>
```

对应 ts 文件,代码如 CORE0215 所示。

代码 CORE0215   ts 文件
```
import { Component } from '@angular/core';
@Component({
  templateUrl: 'hello-ionic.html'
})
export class HelloIonicPage { }
```

3 滑动列表

滑动列表通过向左或向右滑动,以显示或隐藏一组按钮。通过在 <ion-list> 标签内添加 <ion-item-sliding> 和 <ion-item-options> 标签实现滑动列表。其中 <ion-item-sliding> 用于显示列表的内容,<ion-item-options> 显示滑动列表中所对应的按钮。使用滑动列表效果如图 2.13 所示。

图 2.13 滑动列表

为了实现图 2.13 效果,新建界面,代码如 CORE0216 所示。

代码 CORE0216   滑动列表
```
<ion-header>
  <ion-navbar>
    <ion-title> 滑动列表 </ion-title>
```

```
    </ion-navbar>
  </ion-header>
  <ion-content class="outer-content">
    <ion-list>
      <ion-list-header>
        哈哈
      </ion-list-header>
      <ion-item-sliding>
        <ion-item>
          <h2> 小狗 </h2>
          <p> 动物乐园欢乐多 </p>
        </ion-item>
        <ion-item-options>
          <button ion-button color="light" icon-left>
            <ion-icon name="logo-angular"></ion-icon>A
          </button>
          <button ion-button color="primary" icon-left>
            <ion-icon name="ionitron"></ion-icon>B
          </button>
          <button ion-button color="secondary" icon-left>
            <ion-icon name="happy"></ion-icon>C
          </button>
        </ion-item-options>
      </ion-item-sliding>
    <!--省略部分代码 -->
    </ion-list>
  </ion-content>
```

对应 ts 文件，代码如 CORE0217 所示。

代码 CORE0217　ts 文件

```
import { Component } from '@angular/core';
@Component({
  templateUrl: 'hello-ionic.html'
})
export class HelloIonicPage { }
```

技能点 4　Ionic 功能

1　Ionic 卡片

近年来卡片在前端设计使用越来越频繁，卡片具有内容显示分层次、更加规整等特点。在卡片中可以使用列表、图像、标题等组件。卡片的基本结构代码如 CORE0218 所示。

代码 CORE0218　卡片的基本结构

```
<ion-card>
<!-- 卡片头部 -->
    <ion-card-header>
    </ion-card-header>
<!-- 卡片内容 -->
    <ion-card-content>
    </ion-card-content>
</ion-card>
```

使用卡片效果如图 2.14 所示。

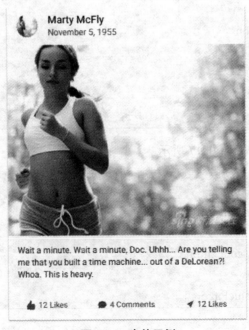

图 2.14　卡片示例

为了实现图 2.14 效果,新建界面,代码如 CORE0219 所示。

代码 CORE0219 卡片

```
<ion-card>
 <ion-item>
  <ion-avatar item-start>
   <img src="assets/img/mike.png">
  </ion-avatar>
  <h2>Marty McFly</h2>
  <p>November 5, 1955</p>
 </ion-item>
 <img src="assets/img/mike.png">
 <ion-card-content>
<!-- 卡片内容 -->
 </ion-card-content>
 <ion-row>
  <ion-col>
   <button ion-button icon-left clear small>
    <ion-icon name="thumbs-up"></ion-icon>
    <div>12 Likes</div>
   </button>
  </ion-col>
 </ion-row>
<!--省略部分代码 -->
</ion-card>
```

2 轮播组件

轮播图的效果是多张图片在一定时间间隔内进行循环播放,其主要作用是吸引用户眼球。它是由多个容器组成,每个容器之间可以滑动切换,其代码结构由轮播图容器(<ion-slides> 标签)和轮播图组件(<ion-slide> 标签)组成。Ionic 轮播图提供多种监听事件,如:(ionSlideDidChange)=" 方法 ()",主要事件如表 2.5 所示。

表 2.5 轮播图事件

事件	说明
ionSlideAutoplay	当幻灯片移动时触发
ionSlideAutoplayStart	自动播放启动时触发
ionSlideAutoplayStop	自动播放停止时触发

事件	说明
ionSlideDidChange	幻灯片更换结束时触发
ionSlideDoubleTap	当用户双击滑块的容器时触发
ionSlideDrag	当幻灯片移动时触发
ionSlideNextEnd	当幻灯片更改以"向前"方向结束时触发
ionSlideReachEnd	当幻灯片到达最后一张幻灯片时触发
ionSlideReachStart	当幻灯片到达其初始位置时触发（初始位置）
ionSlideTap	当用户点击幻灯片的容器时触发
ionSlideWillChange	幻灯片更改开始时触发

轮播图具有多种属性，通过这些属性可以更改轮播图的播放状态，属性如表 2.6 所示。

表 2.6　Ionic 轮播图属性

属性	说明
autoplay	转换之间的延迟（以毫秒为单位）
centeredSlides	在屏幕中央放置一个幻灯片
control	传递另一个 Slides 实例或数组的 Slides 实例，该实例应该由此幻灯片实例控制
direction	滑动方向："水平"或"垂直"
initialSlide	初始幻灯片的索引号。默认值：0
paginationType	分页类型 可能的值有：bullets，fraction，progress。默认值：bullets
slidesPerView	每张幻灯片在同一时间可见。默认值：1
spaceBetween	幻灯片之间的距离（px）默认值：0
speed	幻灯片之间的转换持续时间（以毫秒为单位）。默认值：300
zoom	如果为真，则启用缩放功能

通过调用轮播组件提供的方法，可以进行手动/自动切换到指定的幻灯片等操作，方法如表 2.7 所示。

表 2.7　Ionic 轮播图方法

方法	说明
getActiveIndex()	获取活动幻灯片的索引
slideNext()	转换到下一张幻灯片
slidePrev()	转换到上一张幻灯片
slideTo()	转换到指定的幻灯片
startAutoplay()	开始自动播放

续表

方法	说明
stopAutoplay()	停止自动播放
isEnd()	获取当前幻灯片是否是最后一张幻灯片
length()	获取幻灯片的总数
lockSwipeToNext(should-LockSwipeToNext)	锁定或解锁滑动到下一张幻灯片的能力。shouldLockSwipeToNext: 如果设置为 true,则用户将无法向下滑动。设置为 false 以解锁此行为
update()	更新底层滑块实现。如果已添加或删除子幻灯片,请将其调用

使用轮播图效果如图 2.15 所示。

(a)

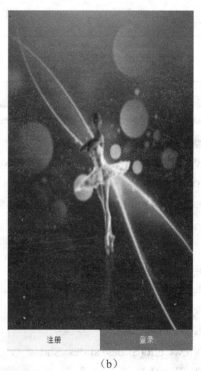
(b)

图 2.15　轮播图

为了实现图 2.15 效果,新建界面,代码如 CORE0220 所示。

代码 CORE0220　轮播图

```
<ion-slides id="containers" #mySlider>
  <ion-slide>
    <img class="box" src="assets/img/1.jpg">
  </ion-slide>
  <ion-slide>
    <img class="box" src="assets/img/2.jpg">
```

```
      </ion-slide>
      <ion-slide>
        <img class="box" src="assets/img/3.jpg">
      </ion-slide>
    </ion-slides>
    <!--ion-button block 为 ionic 预定义样式; -->
    <button ion-button block color="light" (click)="register()" class="regis">
    注册
    </button>
    <button ion-button block color="danger" (click)="login()" class="log">
    登录
    </button>
```

对应 ts 文件,代码如 CORE0221 所示。

代码 CORE0221   ts 文件

```
import { Component } from '@angular/core';
import { Slides } from 'ionic-angular';
import { ViewChild } from '@angular/core';
@Component({
  selector: 'page-home',
  templateUrl: 'home.html'
})
export class HomePage {
  constructor() {}
}
class MyPage {
  @ViewChild(Slides) slides: Slides;
  goToSlide() {
    this.slides.slideTo(2, 200);
  }
}
```

3 数据查询

在 Ionic 项目中,当列表数据过多时,要找一条特定的记录尤为困难,这时数据查询功能就可以用到。其实现需要调用 getItems() 方法,调用该方法的触发类型有三种,具体如表 2.8 所示。

表 2.8　触发类型

类　型	含　义
ionCancel	单击取消按钮时触发
ionClear	单击清除按钮时触发
ionInput	当搜索栏输入更改时触发，包括清除时

使用数据查询的效果如图 2.16 和图 2.17 所示。

图 2.16　数据查询前　　　　　　　　图 2.17　数据查询后

为了实现图 2.16 和图 2.17 效果，新建界面，代码如 CORE0222 所示。

代码 CORE0222　　数据查询

```
<ion-searchbar (ionInput)="getItems($event)"></ion-searchbar>
<ion-list>
  <ion-item *ngFor="let item of items">
    {{ item }}
  </ion-item>
</ion-list>
```

对应 ts 文件，代码如 CORE0223 所示。

代码 CORE0223　　ts 文件

```
import { Component } from '@angular/core';
import { NavController } from 'ionic-angular';
```

```typescript
@Component({
  selector: 'page-home',
  templateUrl: 'home.html'
})
export class HomePage {
  items;
  constructor() {
    this.initializeItems();
  }
  initializeItems() {
    this.items = [
      'Angular JS',
      'Bootstrap',
      'HTML',
      'CSS',
      'JavaScript',
      'Ionic1',
      'Ionic2',
      'JQuery',
      'Ajax',
      'Node js',
      'JSP',
      'PHP',
      'MVC',
    ];
  }
  getItems(ev) {
    this.initializeItems();
    var val = ev.target.value;
    if (val && val.trim() != '') {
      this.items = this.items.filter((item) => {
        return (item.toLowerCase().indexOf(val.toLowerCase()) > -1);
      })
    }
  }
}
```

项目二 "Dancer 时代"首页模块的实现 61

通过下面十二个步骤的操作,实现图 2.2 所示的"Dancer 时代"首页模块及所对应的功能。

第一步:找到项目所在路径,使用"shift+ 鼠标右键"打开"在此处打开命令窗口",输入 Ionic g page tabs 创建首页选项卡,并创建每个选项卡需要显示的界面,如图 2.18 所示。

图 2.18 创建界面

第二步:打开 Sublime Text 2 软件,进行首页选项卡配置,在 app.module.ts 中配置各界面的路径,需要引入选项卡所包含的各个页面。代码如 CORE0224 所示。

代码 CORE0224   app.module.ts

```
import { NgModule, ErrorHandler } from '@angular/core';
import { BrowserModule } from '@angular/platform-browser';
import { IonicApp, IonicModule, IonicErrorHandler } from 'Ionic-angular';
import { MyApp } from './app.component';

import { TabsPage } from '../pages/tabs/tabs';
import { FriendPage } from '../pages/friend/friend';
import { MinePage } from '../pages/mine/mine';
import { HomePage } from '../pages/home/home';
import { Core0101Page } from '../pages/CORE0101/CORE0101';
import { LoginPage } from '../pages/login/login';
import { AddPage } from '../pages/add/add';

import { StatusBar } from '@Ionic-native/status-bar';
```

```typescript
import { SplashScreen } from '@Ionic-native/splash-screen';
// 需要配置的 ts 路径
@NgModule({
  declarations: [
    MyApp,
    FriendPage,
    MinePage,
    HomePage,
    TabsPage,
    LoginPage,
    AddPage,
    Core0101Page // 必须引入的参数名
  ],
  imports: [
    BrowserModule,
    IonicModule.forRoot(MyApp,{
      tabsHideOnSubPages: 'true'      // 隐藏全部子页面 tabs
    })
  ],
  bootstrap: [IonicApp],
  entryComponents: [
    MyApp,
    FriendPage,
    MinePage,
    HomePage,
    TabsPage,
    LoginPage,
    AddPage,
    Core0101Page // 必须引入的参数名
  ],
  providers: [
    StatusBar,
    SplashScreen,
    {provide: ErrorHandler, useClass: IonicErrorHandler}
  ]
})
export class AppModule {}
```

第三步：在 tabs.ts 中引入选项卡所包含的各个页面并且进行相应界面的设置，代码如 CORE0225 所示。

代码 CORE0225　tab.ts

```typescript
import { Component } from '@angular/core';
import { FriendPage } from '../friend/friend';
// 导入分享界面
import { MinePage } from '../mine/mine';
// 导入我的界面
import { HomePage } from '../home/home';
// 导入首页
import { AddPage } from '../add/add';
// 导入上传界面
@Component({
  templateUrl: 'tabs.html'
})
export class TabsPage {
  tab1Root = HomePage;
  // 设置选项卡对应界面
  tab2Root = AddPage;
  tab3Root = FriendPage;
  tab4Root = MinePage;
  constructor() {

  }
}
```

第四步：将 tabs.ts 中每个界面对应的选项卡变量填入相应的 [root] 中，代码如 CORE0226 所示。

代码 CORE0226　tab.html

```html
<ion-tabs>
  <!--[root] 对应 ts 中的名称 -->
  <ion-tab [root]="tab1Root"></ion-tab>
  <ion-tab [root]="tab2Root"></ion-tab>
  <ion-tab [root]="tab3Root"></ion-tab>
  <ion-tab [root]="tab4Root"></ion-tab>
</ion-tabs>
```

第五步：在 home.ts、friend.ts、add.ts、mine.ts 配置 class 类的名称（跟 app.module.ts 中设置的名称一样），部分代码如 CORE0227 所示。

代码 CORE0227　home.ts

```typescript
import { Component } from '@angular/core';
import { NavController, NavParams } from 'Ionic-angular';
@Component({
  selector: 'page-home',
  templateUrl: 'home.html',
})
export class HomePage {
  constructor(public navCtrl: NavController, public navParams: NavParams) {
  }
}
```

第六步：选项卡样式的制作。

底部选项卡由文字和图标组成，通过 tabTitle 和 tabIcon=" 图标名 " 设置文字和图标。代码如 CORE0228 所示。效果如图 2.19 所示。

代码 CORE0228　导航条部分

```html
<ion-tabs>
  <!--tabIcon 为选项卡图标的设置后边是图标的名称 -->
  <!--tabTitle 为选项卡文字的设置 -->
  <!--[root] 对应 ts 中的名称 -->
  <ion-tab [root]="tab1Root" tabTitle=" 首页 " tabIcon="home"></ion-tab>
  <ion-tab [root]="tab2Root" tabTitle=" 上传 " tabIcon="cloud-upload"></ion-tab>
  <ion-tab [root]="tab3Root" tabTitle=" 分享 " tabIcon="redo"></ion-tab>
  <ion-tab [root]="tab4Root" tabTitle=" 我的 " tabIcon="person"></ion-tab>
</ion-tabs>
```

项目二 "Dancer 时代"首页模块的实现

图 2.19 选项卡效果图

第七步:项目首页轮播图的制作。

轮播图采用轮播组件(slides)制作,代码 CORE0229 如下。设置样式前效果如图 2.20 所示。

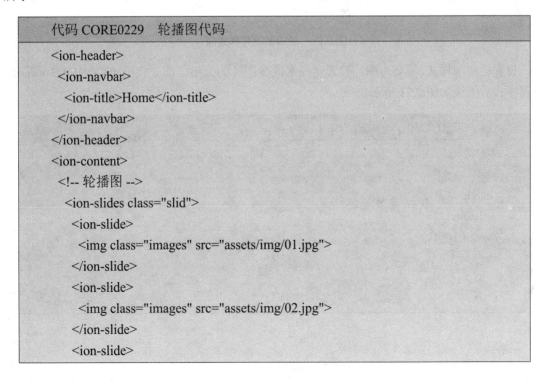

代码 CORE0229　轮播图代码

```
<ion-header>
 <ion-navbar>
  <ion-title>Home</ion-title>
 </ion-navbar>
</ion-header>
<ion-content>
 <!-- 轮播图 -->
 <ion-slides class="slid">
  <ion-slide>
   <img class="images" src="assets/img/01.jpg">
  </ion-slide>
  <ion-slide>
   <img class="images" src="assets/img/02.jpg">
  </ion-slide>
  <ion-slide>
```

```
        <img class="images" src="assets/img/03.jpg">
      </ion-slide>
    </ion-slides>
</ion-content>
```

图 2.20　轮播图设置样式前

设置轮播图样式，需要为图片设置宽高来改变图片的大小。部分代码如 CORE0230 所示。设置样式后效果如图 2.21 所示。

代码 CORE0230　轮播图 SCSS 代码

```scss
// 轮播图样式
.slid{
  height: 43%;
  .images{
    width:100%;
    height:100%;
  }
}
```

图 2.21 轮播图设置样式后

第八步：项目首页图标区域的制作。

图标区域是由图标和文字组成。部分代码如 CORE0231 所示。效果如图 2.22 所示。

```
代码 CORE0231    图标区域 HTML
    <!-- 轮播图下方文字和图标的组合 -->
    <div class="row">
      <div class="col">
        <ion-icon name="wine" class="icon street"></ion-icon><br>
        <span class="category"> 街舞 </span>
      </div>
      <div class="col">
        <ion-icon name="beer" class="icon jazz"></ion-icon><br>
        <span class="category"> 爵士舞 </span>
      </div>
      <div class="col">
        <ion-icon name="trophy" class="icon modern"></ion-icon><br>
        <span class="category"> 现代舞 </span>
      </div>
    <div class="col">
```

```
            <ion-icon name="cafe" class="icon folk"></ion-icon><br>
    <span class="category"> 民族舞 </span>
      </div>
    </div>
```

图 2.22　图标区域设置样式后

设置图标区域样式，需要设置图标和文字的大小和位置。部分代码如 CORE0232 所示。设置样式后效果如图 2.23 所示。

代码 CORE0232　图标区域 scss 代码

```
// 轮播图下方模块的大小
.row{
  height: 12%;
    text-align: center;
  line-height: 150%;
}
// 模块中图标的样式
.icon{
  font-size: 200%;
}
```

```
// 分类的图标颜色
.street{
  color: yellow;
}
.jazz{
  color: #007aff;
}
.modern{
  color: #f53d3d;
}
.folk{
  color: #2ec95c;
}
```

图 2.23　图标区域设置样式后

第九步：首页列表的制作

列表是在 <ion-item> 标签里面嵌入图片文字等数据。部分代码如 CORE0233 所示。效果如图 2.24 所示。

代码 CORE0233　　列表 html

```html
<hr id="line">
<!-- 视频介绍 -->
  <ion-item (click)="videolist()">
    <div class="media">
      精选视频 <br><br>
      <div>
        <img src="assets/img/001.jpg" alt="...">
      </div>
      <div>
        <h4> 舞蹈是一种表演艺术,使用身体来完成各种优雅或高难度的动作,一般有音乐伴奏,以有节奏的动作为主要表现手段的艺术形式。</h4>
      </div>
    </div>
  </ion-item>
<!-- 音乐介绍 -->
  <ion-item (click)="musicsing()">
    <div class="media">
      精选音乐 <br><br>
      <div>
        <img src="assets/img/002.jpg" alt="...">
      </div>
      <div>
        <h4> 舞蹈是一种表演艺术,使用身体来完成各种优雅或高难度的动作,一般有音乐伴奏,以有节奏的动作为主要表现手段的艺术形式。</h4>
      </div>
    </div>
  </ion-item>
```

设置列表样式,需要设置文字的大小和位置,图片的大小以及图片下方文字的省略。部分代码如 CORE0234 所示。设置样式后效果如图 2.25 所示。

项目二 "Dancer 时代"首页模块的实现

图 2.24 列表设置样式前

代码 CORE0234　列表 scss 代码

// 两个模块之间的间距
#line{
　height: 2%;
}
// 视频音乐模块的样式设置
.media{
　width: 100%;
}
// 图片的样式
.media img{
　width: 100%;
　height: 180px;
}
// 字体的样式
.media h4{
　text-overflow: ellipsis;
　overflow: hidden;
　white-space: nowrap;
　width: 95%;
}

图 2.25 列表设置样式后

第十步：创建音乐列表界面并配置。

第十一步：创建视频列表界面并配置。

第十二步：首页跳转的实现，当点击精选视频或者精选音乐时发生跳转，进入视频界面或者音乐界面，代码如 CORE0235 所示。

代码 CORE0235　首页面跳转（home.ts）

import { Component } from '@angular/core';

import { NavController, NavParams } from 'Ionic-angular';

import { VideolistPage } from '../videolist/videolist';
// 导入视频分类列表界面

import { MusicsingPage } from '../musicsing/musicsing';
// 导入音乐界面

@Component({

　selector: 'page-home',

　templateUrl: 'home.html',

})

export class HomePage {

　constructor(public navCtrl: NavController, public navParams: NavParams) {

　}

```
videolist(){
  this.navCtrl.push(VideolistPage);
  // 跳转到视频分类界面
}
musicsing(){
  this.navCtrl.push(MusicsingPage);
  // 跳转到音乐界面
}
}
```

至此,"Dancer 时代"首页模块制作完成。

本项目通过对"Dancer 时代"首页模块的学习,对 Ionic 相关组件及应用列表布局的使用具有初步了解,对 Ionic 相关的导航及菜单栏的使用有所认识,同时掌握了 Ionic 使用组件实现首页的布局和美化。

component	组件	toggle	切换
placeholder	占位符	checked	选中的
layout	布局	outline	轮廓
disabled	禁用	round	圆形的
full	完全的		

一、选择题

1. Placement 属性可以设置 Tabs(　　),属性值分别有:top、bottom,默认值为 bottom。
A. 出现的位置　　B. 是否高亮选中效果　　C. 添加一个图标　　D. 添加图标和文字
2. (　　)用于设置 ionNavBar 中按钮放置的位置。可以使用的属性值有:left 或 right。
A.side　　　　B.ion-arrow　　　　C.ionNavBar　　　　D.ionNavView
3. 在 Ionic 3 中,使用(　　)指令定义侧边栏菜单。
A.ion-side-menu　　B.ion-side-menus　　C.ion-side-menu-conten　　D.ion-side-conten
4. 创建和配置轮播图后,使用(　　)方法可以导航到特定的轮播图。

A.slideNext()　　　　B.slide()　　　　　　C.slideTo()　　　　　　D.Next()
5．轮播组件模板包括轮播图容器（　　）和任何数量的轮播图组件。
A.ionSlideDidChange　　　　　　　　B.ViewChild
C.ion-slide　　　　　　　　　　　　D.ion-slides

二、填空题

1．选项卡模块使用＿＿＿＿＿＿标签来定义，选项卡里面的选项使用 ion-tab 标签来定义。
2．在 IOS 系统中，选项卡默认出现在屏幕的底部，Android 系统中默认出现在＿＿＿＿＿＿。
3．ion-view 是＿＿＿＿＿＿的子标签。ion-view 是一个容器，用于展示视图或导航栏信息。
4．＿＿＿＿＿＿是侧边栏打开关闭的切换指令。
5．页面跳转的实现方法有两种，分别是＿＿＿＿＿＿跳转和通过带参数跳转。

三、上机题

编写符合以下要求的网页，实现跳转到选项卡页面的效果。要求：
1．使用 Ionic 知识创建选项卡界面；
2．再创建一个页面并实现跳转，效果如下图。

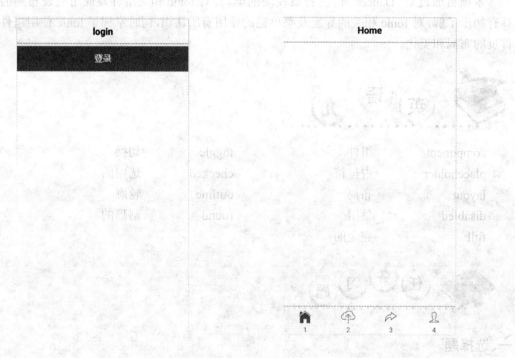

项目三 "Dancer 时代"音频模块的实现

通过"Dancer 时代"音频模块的实现，了解视频、音频播放功能的实现流程，学习 Ionic 弹出框的使用，掌握 Ionic 中插件的安装和使用，具有独立使用插件实现播放功能的能力。在任务实现过程中：

● 了解视频、音频播放功能的实现流程。
● 掌握 Ionic 弹出框的使用。
● 掌握 Ionic 手势事件的实际应用。
● 具有独立使用插件实现播放功能的能力。

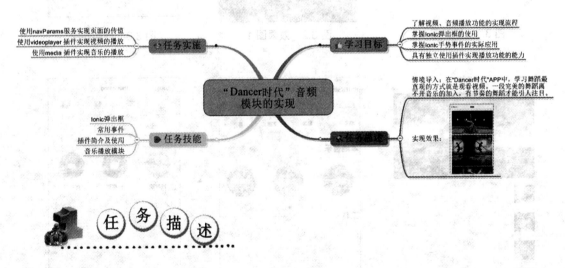

【情境导入】

在"Dancer 时代"APP 中，学习舞蹈最直观的方式就是观看视频。一段完美的舞蹈离不开音乐的加入，有节奏的舞蹈才能引人注目。因此，Ellison 综合各个方面的需求在首页中添加了播放视频、音乐的入口，通过点击进入视频界面，之后是推荐的精彩视频列表，用户可以通过翻阅列表选择自己喜欢的视频进行观看。点击进入音乐界面后,用户可以根据自己的喜好选择音乐。本项目主要通过实现"Dancer 时代"的视频、音频播放功能了解 Ionic 弹出框和插件的使用。

【功能描述】

本项目将实现"Dancer 时代"音频模块的设计及播放。
- 使用 navParams 服务实现跳转页面的传值。
- 使用 videoplayer 插件实现视频的播放。
- 使用 media 插件实现音乐的播放。

【基本框架】

基本框架如图 3.1、图 3.3、图 3.5 所示,通过本项目的学习,能将框架图 3.1、图 3.3、图 3.5 转换成效果图 3.2、图 3.4、图 3.6。

图 3.1 框架图 1

图 3.2 效果图 1

图 3.3 框架图 2

图 3.4 效果图 2

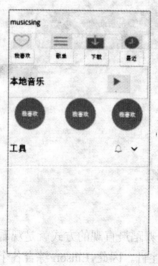

图 3.5 框架图 3

图 3.6 效果图 3

技能点 1 Ionic 弹出框

1 模态框

模态框是常见的对话框组件,覆盖在整个屏幕,常用来进行选择或信息编辑。在模态框关闭之前,其他的用户交互行为将被阻止,只有完成指定操作后,才能返回到原来页面。其实现步骤如下所示。

第一步:在 HTML 页面中定义一个按钮,并添加点击事件。

```
<button ion-button block color="dark" (click)="showAlert()"> 模态框 </button>
```

第二步:在对应的 ts 文件引入 ModalController。

```
import {ModalController } from 'ionic-angular';
```

第三步:构造 ModalController。

```
constructor(public modalCtrl: ModalController){}
```

第四步:通过 showAlert() 实现模态框的功能。

```
showAlert(){
  let modal = this.modalCtrl.create(CallModalPage);
  modal.present();
}
```

在模态框被呈现之后,数据可以通过 Modal.create() 方法传递给弹出的模态框。其 create(component, data, opts) 的参数详解如表 3.1 所示。

表 3.1 create 方法参数

参数	类型	描述
component	object	模态框的视图
data	object	任何传递到模态视图的数据
opts	object	模态框的选项

使用模态框效果如图 3.7 所示。

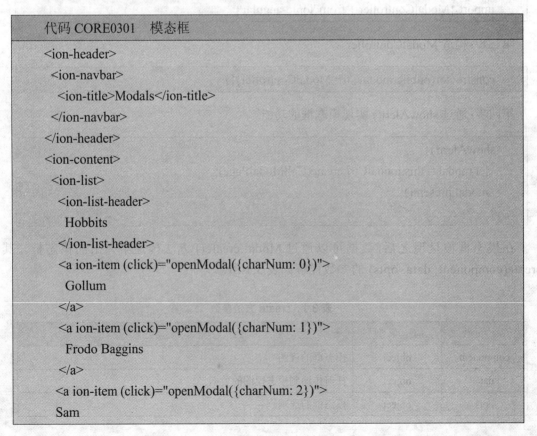

图 3.7 模态框

为了实现图 3.7 的效果,新建界面,代码如 CORE0301 所示。

代码 CORE0301 模态框

```
<ion-header>
  <ion-navbar>
    <ion-title>Modals</ion-title>
  </ion-navbar>
</ion-header>
<ion-content>
  <ion-list>
    <ion-list-header>
      Hobbits
    </ion-list-header>
    <a ion-item (click)="openModal({charNum: 0})">
      Gollum
    </a>
    <a ion-item (click)="openModal({charNum: 1})">
      Frodo Baggins
    </a>
    <a ion-item (click)="openModal({charNum: 2})">
      Sam
```

```
    </a>
   </ion-list>
</ion-content>
```

对应 ts 文件，代码如 CORE0302 所示。

代码 CORE0302　ts 文件

```
import { Component } from '@angular/core';
import { ModalController, Platform, NavParams, ViewController } from 'ionic-angular';
@Component({…})
export class HelloIonicPage {
 constructor(public modalCtrl: ModalController) { }
 openModal(characterNum) {
   let modal = this.modalCtrl.create(ModalContentPage, characterNum);
   modal.present();
 }
}
@Component({
 template: `
// 省略部分代码
<ion-content>
  <ion-list>
    <ion-item>
      <ion-avatar item-start>
        <img src="{{character.image}}">
      </ion-avatar>
      <h2>{{character.name}}</h2>
      <p>{{character.quote}}</p>
    </ion-item>
    <ion-item *ngFor="let item of character['items']">
      {{item.title}}
      <ion-note item-end>
        {{item.note}}
      </ion-note>
    </ion-item>
  </ion-list>
`
</ion-content>
})
```

```
export class ModalContentPage {
  character;
  constructor(
    public platform: Platform,
    public params: NavParams,
    public viewCtrl: ViewController
  ) {
    var characters = [
      {
        name: 'Gollum',
        quote: 'Sneaky little hobbitses!',
        image: 'assets/img/avatar-gollum.jpg',
        items: [
          { title: 'Race', note: 'Hobbit' },
          { title: 'Culture', note: 'River Folk' },
          { title: 'Alter Ego', note: 'Smeagol' }
        ]
      },
// 省略部分代码
    this.character = characters[this.params.get('charNum')];
  }
  dismiss() {
    this.viewCtrl.dismiss();
  }
}
```

2 对话框

对话框（Alerts）通常用于给用户传达相关提醒信息，在用户回应之前不可以进行其他操作。对话框与模态框不同，对话框仅占据部分屏幕空间。意味着对话框只能用来做一些快速的信息交互，比如密码确认、小提示等。其实现步骤如下所示。

第一步：在 HTML 页面中定义一个按钮，并添加点击事件。

```
<button ion-button block color="dark" (click)="showAlert()"> 对话框 </button>
```

第二步：在对应的 ts 文件引入 AlertController。

```
import { AlertController } from 'ionic-angular';
```

第三步：构造 AlertController。

```
constructor(public alertCtrl: AlertController){}
```

第四步：通过 showAlert() 实现对话框的功能。

```
showAlert() {
  let alert = this.alertCtrl.create({
    // 对话框参数
  });
  alert.present();
}
```

对话框分类如表 3.2 所示。

表 3.2 对话框分类

类　型	描　　述
基本弹窗	包含一个按钮供关闭弹出框
带输入的弹窗	包含一个文本输入框、一个取消按钮和确认按钮
确认框	包含一个取消按钮和确认按钮
带单选的弹窗	包含一个单选组件、取消和确认按钮，是 radio 组件和弹窗组件的结合
带多选的弹窗	包含一个多选组件、取消和确认按钮。带多选的弹框是 checkbox 组件和弹窗组件的结合

使用对话框效果如图 3.8 所示。

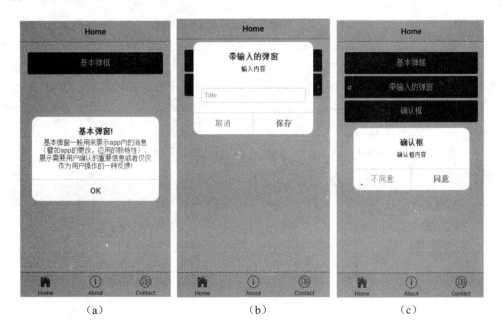

　　　　（a）　　　　　　　　　　　（b）　　　　　　　　　　　（c）

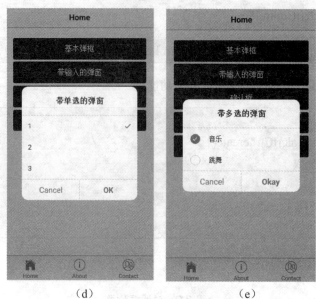

(d)　　　　　　　　　(e)

图 3.8　对话框示例

为了实现图 3.8 的效果,新建界面,代码如 CORE0303 所示。

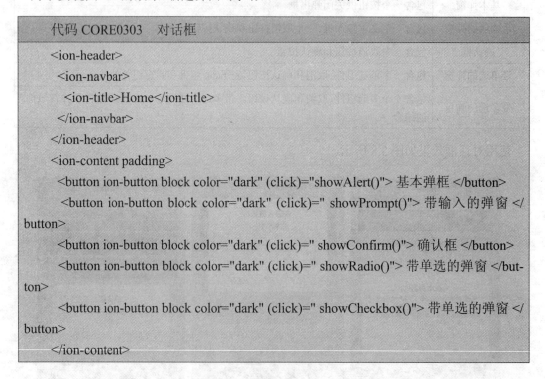

对应 ts 文件，代码如 CORE0304 所示。

代码 CORE0304　ts 文件

```ts
import { Component } from '@angular/core';
import { NavController } from 'ionic-angular';
// 引入 AlertController
import { AlertController } from 'ionic-angular';
@Component({
  selector: 'page-home',
  templateUrl: 'home.html'
})
export class HomePage {
  testCheckboxOpen: boolean;
  testCheckboxResult;
  testRadioOpen: boolean;
  testRadioResult;
  // 声明 AlertController
constructor(public navCtrl: NavController,public alertCtrl: AlertController) { }
// 基本弹窗方法
showAlert() {
  let alert = this.alertCtrl.create({
    title: ' 基本弹窗 !',
    subTitle: ' 基本弹窗一般用来展示 app 内的消……!',
 buttons: ['OK']
  });
  alert.present();
}
// 带输入弹窗方法
  showPrompt() {
    let prompt = this.alertCtrl.create({
      title: ' 带输入的弹窗 ',
      message: " 输入内容 ",
      inputs: [
        {
          name: 'title',
          placeholder: 'Title'
        },
      ],
```

```
      buttons: [
        {
          text: '取消',
          handler: data => {
            console.log('Cancel clicked');
          }
        },
        {
          text: '保存',
          handler: data => {
            console.log('Saved clicked');
          }
        }
      ]
    });
    prompt.present();
  }
// 确认弹窗方法
  showConfirm() {
    let confirm = this.alertCtrl.create({
    title: '确认框',
      message: '确认框内容',
      buttons: [
        {
          text: '不同意',
          handler: () => {
            console.log('Disagree clicked');
          }
        },
        {
          text: '同意',
          handler: () => {
            console.log('Agree clicked');
          }
        }
      ]
    });
    confirm.present();
```

```
  }
// 带单选按钮弹窗方法
  showRadio() {
    let alert = this.alertCtrl.create();
    alert.setTitle(' 带单选的弹窗 ');
      alert.addInput({
      type: 'radio',
      label: '1',
      value: '1',
      checked: true
    });
    alert.addInput({
      type: 'radio',
      label: '2',
      value: '2'
    });
    alert.addInput({
      type: 'radio',
      label: '3',
      value: '3'
    });
     alert.addButton('Cancel');
    alert.addButton({
      text: 'OK',
      handler: data => {
        this.testRadioOpen = false;
        this.testRadioResult = data;
      }
    });
    alert.present();
  }
// 带多选按钮弹窗方法
  showCheckbox() {
    let alert = this.alertCtrl.create();
    alert.setTitle(' 带多选的弹窗 ');
// 带多选按钮弹窗的选项
    alert.addInput({
      type: 'checkbox',
```

```
      label: '音乐',
      value: 'value1',
      checked: true
    });
    alert.addInput({
      type: 'checkbox',
      label: '跳舞',
      value: 'value2'
    });
    alert.addButton('Cancel');
    alert.addButton({
      text: 'Okay',
      handler: data => {
        console.log('Checkbox data:', data);
    this.testCheckboxOpen = false;
        this.testCheckboxResult = data;
      }
    });
    alert.present();
  }
}
```

3 上拉菜单

上拉菜单（Action Sheets）是从设备屏幕的底部边缘向上滑出的弹出框。其内容通常以列表的形式显示在页面的最下方,可通过点击其所在页面使其消失。当被触发时,其所在页面将会变暗,信息无法修改。其实现步骤如下所示。

第一步:在 HTML 页面中定义一个按钮,并添加点击事件。

```
<button class="add-button" ion-button color="light" round (click)="doConfirm()">上传</button>
```

第二步:在对应的 ts 文件引入 ActionSheetController。

```
import { ActionSheetController } from 'ionic-angular';
```

第三步:构造 ActionSheetController。

```
constructor(public actionSheetCtrl: ActionSheetController) {}
```

第四步：通过 doConfirm() 实现上拉菜单的功能。

```
doConfirm() {
  let actionSheet = this.actionSheetCtrl.create({
    title: ' 文件上传 ',
    // 上拉菜单按钮组
    buttons: [
      {
        text: ' 选择文件 ',
        handler: () => {
          alert(' 选择文件 ');
        }
      },
    ]
  });
  actionSheet.present();
}
```

上拉菜单是由一个按钮组组成，每个按钮包括 text 属性、可选的 handler 和 role 属性。按钮的 role 属性可以是 destructive（始终是阵列中的第一个按钮）或 cancel（始终作为底部按钮）。没有 role 属性的按钮将具有该平台的默认顺序，所有其他按钮将按照已添加到 buttons 阵列的顺序显示。上拉菜单属性具体如表 3.3 所示。

表 3.3　上拉菜单属性

属　　性	类　　型	描　　述
title	string	上拉菜单按钮的标题
Icon	Icon	上拉菜单按钮的图标
cssClass	string	附加的 CSS 样式类，简化的特定弹出框
handler	any	是一个 function，用来处理该按钮的 click 事件
role	string	按钮显示的顺序，有可选的 destructive 或 cancel 属性。没有角色属性的按钮将具有该平台的默认顺序

使用上拉菜单效果如图 3.9 所示。

图 3.9 上拉菜单

为了实现图 3.9 的效果,新建界面,代码如 CORE0305 所示。

代码 CORE0305　上拉菜单

```
<ion-header>
  <ion-navbar>
    <ion-title>add</ion-title>
// 绑定点击事件触发按钮
    <button class="add-button" ion-button color="light" round (click)="doConfirm()">上传 </button>
  </ion-navbar>
</ion-header>
<ion-content padding>
<!--省略部分代码 -->
</ion-content>
```

对应 ts 文件,代码如 CORE0306 所示。

代码 CORE0306　ts 文件

```
import { ActionSheetController } from 'ionic-angular';
export class HomePage {
  constructor(public actionSheetCtrl: ActionSheetController) {}
// 对应的方法
```

```
  doConfirm() {
    let actionSheet = this.actionSheetCtrl.create({
      title: ' 文件上传 ',
// 上拉菜单按钮组
      buttons: [
        {
          text: ' 选择文件 ',
          handler: () => {
            alert(' 选择文件 ');
          }
        },
        {
          text: ' 上传 ',
          handler: () => {
            alert(' 上传 ');
          }
        },
      {
          text: ' 取消 ',
          role: 'cancel',
          handler: () => {
            alert(' 取消 ');
          }
        }
      ]
    });
    actionSheet.present();
  }
}
```

技能点 2　常用事件

1　触发事件

触屏是身边电子设备的常态,手势事件随着触屏的出现被越来越多的用户所接受。使用手势事件能够给用户提供良好的交互特性,从而使用户更方便、快捷地进行操作。Ionic 框架支持多种手势事件,其主要手势事件如下表 3.4 所示。

表 3.4　Ionic 支持的手势事件

事　件	描　　述
tap	手指在元素上点击后立即触发
press	手指在元素上按下后立即触发
pan	手指在元素上滑动触发
swipe	手指在元素上迅速滑动触发
rotate	手指在元素上旋转触发
pinch	手指在元素上缩放触发

定义手势事件的响应非常直接,以点击事件为例,tap 与 click 都会触发点击事件,其中 tap 事件主要用于移动端,click 事件主要用于 PC 端,如果想用于移动端,则会比 tap 事件延迟 200~300 ms。使用 tap 代码如下所示。

```
<ion-card (tap)="tapEvent($event)">
  <ion-item>
    点击：{{tap}} 次
  </ion-item>
</ion-card>
```

使用手势效果如图 3.10 所示。

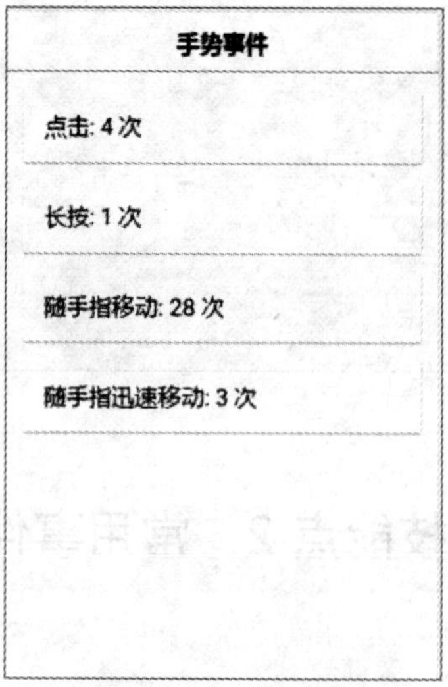

图 3.10　手势事件

为了实现图 3.10 效果，新建界面，代码如 CORE0307 所示。

代码 CORE0307　手势事件

```html
<ion-header>
  <ion-navbar>
    <ion-title> 手势事件 </ion-title>
  </ion-navbar>
</ion-header>
<ion-content>
  <ion-card (tap)="tapEvent($event)">
    <ion-item>
      点击：{{tap}} 次
    </ion-item>
  </ion-card>
  <!--省略部分代码 -->
</ion-content>
```

对应 ts 文件，代码如 CORE0308 所示。

代码 CORE0308　ts 文件

```typescript
import { Component } from '@angular/core';
@Component({
  selector: 'page-hello-ionic',
  templateUrl: 'hello-ionic.html'
})
// 默认手指事件次数
export class HelloIonicPage {
  public press: number = 0;
  public pan: number = 0;
  public swipe: number = 0;
  public tap: number = 0;
  constructor() {
  }
  // 随着手指移动次数增加
  pressEvent(e) {
    this.press++
  }
  panEvent(e) {
    this.pan++
```

```
    }
    swipeEvent(e) {
      this.swipe++
    }
    tapEvent(e) {
      this.tap++
    }
  }
```

2　生命周期事件

页面的生命周期一般指从请求页面到完全加载页面的过程,在每个阶段都会执行一些特定的事件来完成指定任务,当前页面的离开、跳转、加载就会出现页面的生命周期事件。Ionic 支持的生命周期事件如表 3.5 所示。

表 3.5　Ionic 支持的生命周期事件

事件	描述
ionViewDidLoad	页面加载时运行。此事件仅在每个页面创建时发生一次
ionViewWillEnter	当页面即将进入并成为活动页面时运行
ionViewDidEnter	当页面已完全输入并且现在是活动页面时运行
ionViewWillLeave	当页面即将离开并且不再是活动页面时运行
ionViewDidLeave	当页面完成离开时运行,并且不再是活动页面
ionViewWillUnload	当页面即将被销毁并删除其元素时运行
ionViewCanEnter	在视图进入之前运行
ionViewCanLeave	运行之前视图可以离开

使用生命周期事件效果如图 3.11 所示。

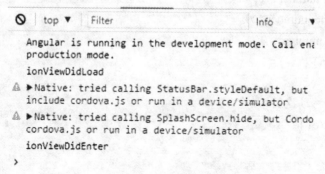

图 3.11　生命周期事件

为了实现图 3.11 的效果,新建界面,对应 ts 文件代码如 CORE0309 所示。

代码 CORE0309　　生命周期

```
import { Component } from '@angular/core';
@Component({
 templateUrl: 'hello-ionic.html'
})
export class HelloIonicPage {
// 调用页面生命周期事件
 ionViewDidLoad() {
  console.log("ionViewDidLoad");
 }
 ionViewDidEnter() {
  console.log("ionViewDidEnter");
 }
}
```

提示　在应用访问多个页面后按返回键，Ionic 会按照页面访问的顺序逐一返回每个访问过的页面，遇到这样的体验，最想做的就是砸掉手机。扫描图中二维码，你的想法是否有所改变呢？

技能点 3　插件简介及使用

1　Cordova 插件

当使用 Ionic 框架制作 APP 时，可通过 Cordova 提供的一组 API 访问原生设备功能，如摄像头拍照、麦克风、地理定位等。其支持 IOS、Android 等移动操作系统，目前提供的 Cordova 插件数有上千个，其官网如图 3.12 所示（http://ionicframework.com/docs/native/）。

图 3.12　Ionic 官网

使用 Cordova 或 Ionic CLI 安装插件，安装的插件放置在 plugins 文件夹中。例如，如果要安装 Camera 插件，则需要运行以下命令。

```
ionic cordova plugin add cordova-plugin-camera
```

使用插件需要安装 Ionic Native 包。Ionic Native 是 Cordova / PhoneGap 插件的一个 TypeScript 包装器。

```
npm install --save @ionic-native/camera
```

Cordova 提供了许多插件用于项目的开发中，大部分插件可以在项目中实现一些非常重要的功能。安装完插件后可以使用以下命令进行管理，以 cordova-plugin-camera 插件为例。
- ionic cordova plugin ls 命令查看所有已经安装的插件。
- ionic cordova plugin add cordova-plugin-camera 命令安装所需插件。
- ionic cordova plugin rm cordova-plugin-camera 命令删除已安装的插件。
- ionic cordova plugin update 更新已安装的插件。

Cordova 插件介绍如表 3.6 所示。

表 3.6　Cordova 插件

名　称	描　述
Network	判断网络连接类型
Device	获取设备信息
Vibration	设备振动
Dialogs	可视化消息提醒
Geolocation	判断设备的物理位置
Media Capture	获取图像、录视频、录音
Media	记录或播放媒体文件
Barcode Scanner	通过摄像头识别二维码/条形码

2　Camera 插件

Camera 插件可以方便地调用系统设备的摄像头拍摄照片和视频，它提供了一个用于拍摄照片和从系统的图像库中选择图像的 API，相机可选参数如下表 3.7 所示。

表 3.7　相机可选参数

参　数	类　型	描　述
quality	Number，默认 50	图像的质量，范围 0~100，其中 100 是通常的全分辨率，而不会损失文件压缩
destinationType	默认 file-url	getPicture 返回值的格式

续表

参数	类型	描述
sourceType	默认 camera	设置图片的来源
encodingType	默认 JPEG	选择返回的图像文件的编码
mediaType	默认 picture	设置要选择的媒体类型
allowEdit	Boolean，默认 true	允许在选择前简单地编辑图像
correctOrientation	Boolean	在拍摄期间旋转图像以校正设备的方向

使用命令窗口安装 Camera 插件代码如下。

```
ionic cordova plugin add cordova-plugin-camera
npm install --save @ionic-native/camera
```

使用 cordovaCamera 插件效果如图 3.13 所示。

为了实现图 3.13 的效果，在 app.module.ts 中引入 Camera 插件，代码如 CORE0310 所示。

图 3.13 Camera 插件

代码 CORE0310　app.module.ts 文件

```
import { Camera, CameraOptions } from '@ionic-native/camera';
// 省略部分代码
providers: [
    StatusBar,
```

```
    SplashScreen,
    Camera,
    {provide: ErrorHandler, useClass: IonicErrorHandler}
  ]
})
```

新建 HTML 文件,代码如下所示。

```html
<ion-header>
 <ion-navbar>
  <ion-title>Home</ion-title>
 </ion-navbar>
</ion-header>
<ion-content padding>
<div class="addspeak-div">
  <textarea class="textarea"></textarea>
  <div class="divpreview">
    <img class="imgpreview" hidden="hidden" src="">
    <span class="videopreview"></span>
  </div>
  <div class="choosediv" (click)="show()">
    <ion-icon id="addspeak-icon" name="camera"></ion-icon><br>
    <span> 照片 / 视频 </span>
  </div>
 </div>
</ion-content>
```

对应 ts 文件,代码如 CORE0311 所示。

代码 CORE0311　ts 文件

```typescript
import { Component } from '@angular/core';
import { NavController,NavParams, ActionSheetController } from 'ionic-angular';
import { Camera, CameraOptions } from '@ionic-native/camera';
@Component({
  selector: 'page-home',
  templateUrl: 'home.html'
})
export class HomePage {
```

```
constructor(public navCtrl: NavController, public navParams: NavParams, public ac-
tionSheetCtrl: ActionSheetController,private camera: Camera) {

  }
  // 调用此方法显示弹出信息
  show(){
    let actionSheet = this.actionSheetCtrl.create({
      buttons: [
        {// 通过相册获取图片
          text: ' 相册 ',
          handler: () => {
            const options: CameraOptions = {
              quality: 100,
              // 输出格式
              destinationType: this.camera.DestinationType.NATIVE_URI,
              // 是否保存到相册
              saveToPhotoAlbum: true,
              // 图片来源 , CAMERA：拍照 ,PHOTOLIBRARY：相册
              sourceType: this.camera.PictureSourceType.PHOTOLIBRARY
            };
            this.camera.getPicture(options).then((imageData) => {
              var ss=document.querySelectorAll(".imgpreview");
              ss[0].removeAttribute("hidden")
              var sss=imageData;
              ss[0].setAttribute("src",sss);
              alert(imageData);
            }, (err) => {
              // Handle error
            });
          }
        },
        {// 通过拍照获取图片
          text: ' 拍照 ',
          handler: () => {
            const options: CameraOptions = {
              quality: 100,
              destinationType: this.camera.DestinationType.NATIVE_URI,
              saveToPhotoAlbum: true,
              sourceType: this.camera.PictureSourceType.CAMERA
```

```
                };

            this.camera.getPicture(options).then((imageData) => {
              var ss=document.querySelectorAll(".imgpreview");
              ss[0].removeAttribute("hidden")
              var sss=imageData;
              ss[0].setAttribute("src",sss);
              alert(imageData);
            }, (err) => {
            });
          }
        },
        {
          text: ' 取消 ',
          role: 'cancel',
          handler: () => {
            console.log('Cancel clicked');
          }
        }
      ]
    });
    actionSheet.present();
  }
}
```

提示 人们常说程序员的生活枯燥,其实这是你不懂程序员。代码以外,这些高智商的人幽默有趣,扫描下方二维码,见识一个不一样的程序"猿"!

技能点 4　音乐播放

音乐是一种艺术,它能反映人们现实生活情感,能让我们心情愉悦,能给人听觉上的享受。因此,音乐在手机中的应用是必不可少的,如:闹钟,来电铃声等,这些音乐的应用需要播放功能的支持。在使用之前需要安装插件。在项目根目录下打开命令窗口,输入命令,安装音乐播放插件。命令如下所示。

项目三 "Dancer 时代"音频模块的实现

```
ionic cordova plugin add cordova-plugin-media
npm install --save @ionic-native/media
```

安装后需要将插件引入项目并进行配置。在 app.module.ts 文件中引入插件,代码如下所示。

```
import { Media, MediaObject } from '@ionic-native/media';
//...
imports: [
  IonicModule.forRoot(MyApp),
  Media,
],
```

在需要音乐播放的界面设置播放、暂停等按钮,用于播放和暂停音乐。代码如下所示。

```html
<div class="startmodel">
    <span class="upmusic1" style="float: left;">
    <ion-icon class="upmusic-icon" name="rewind"></ion-icon>
</span>
     <span class="onload">
    <ion-icon class="iconCutPlay" name="play" (click)="playmusic()"></ion-icon>
    <!-- 播放按钮 -->
    <ion-icon class="iconCutPause" hidden name="pause" (click)="pausemusic()">
</ion-icon>
    <!-- 暂停按钮 -->
    </span>
     <span class="upmusic">
       <ion-icon class="upmusic-icon" name="fastforward"></ion-icon>
     </span>
   </div>
```

在对应界面的 ts 文件中添加实现方法。代码如下所示。

```
// 定义音乐对象
  file:MediaObject;
  playmusic() {
// 设置音乐路径
this.file=this.media.create('http://so1.111ttt.com:8282/2017/1/05m/09/298092040183.m4a?tflag=1496216227&pin=58e48cea3b46ba58686a112cc6df0e73&ip=111.165.165.102#.mp3');
```

```
    this.file.play();
    // 播放音乐
    document.querySelectorAll(".iconCutPlay")[0].setAttribute("hidden","hidden");
    // 隐藏播放按钮
    document.querySelectorAll(".iconCutPause")[0].removeAttribute("hidden");
    // 显示暂停按钮
  }
  pausemusic(){
    this.file.pause();
    // 暂停音乐
    document.querySelectorAll(".iconCutPlay")[0].removeAttribute("hidden");
    // 显示播放按钮
    document.querySelectorAll(".iconCutPause")[0].setAttribute("hidden","hidden");
    // 隐藏暂停按钮
  }
```

音乐播放常用方法如表 3.8 所示。

表 3.8 常用方法

方 法	描 述
create(src,onStatusUpdate,onSuccess,onError)	src：音频内容的 URL，onStatusUpdate：当文件的状态更改为可选时，将调用回调函数，onSuccess：完成当前播放，录制或停止动作后调用的回调函数，onError：如果发生错误，将调用回调函数
getCurrentAmplitude()	获取当前录音的当前幅度
getCurrentPosition()	获取音频文件中的当前位置
pause()	暂停播放音频文件
release()	释放底层操作系统的音频资源

使用音乐播放效果如图 3.14 所示。

图 3.14 音乐播放

为了实现图 3.14 的效果,新建界面,代码如 CORE0312 所示。

代码 CORE0312　HTML 文件

```
<ion-header>
 <ion-navbar>
<ion-title> 音乐播放 </ion-title>
 </ion-navbar>
</ion-header>
<ion-content>
   <img src="assets/img/001.jpg" style="width: 80%;margin-left: 10%;height: 85%;margin-top: 10%;">
   <!-- 图片区域 -->
</ion-content>
<ion-footer>
  <div class="audio-Cut">
  <div class="Cut-div1">
     <img class="musicimage" src="assets/img/001.jpg" >
     <!-- 底部右侧显示的图片 -->
  </div>
    <div class="Cut-div2">
    <div class="Cut-div3">
      <p class="footmusicname"> 凉城 </p>
```

```html
        <p class="footsingername"> 任然 </p>
        <!-- 作者和歌曲名称 -->
    </div>
    <div class="startmodel">
        <span class="upmusic1" style="float: left;">
            <ion-icon class="upmusic-icon" name="rewind"></ion-icon>
        </span>
        <span class="onload">
    <ion-icon class="iconCutPlay" name="play" (click)="playmusic()"></ion-icon>
            <!-- 播放按钮 -->
    <ion-icon class="iconCutPause" hidden name="pause" (click)="pausemusic()">
    </ion-icon>
            <!-- 暂停按钮 -->
        </span>
        <span class="upmusic">
    <ion-icon class="upmusic-icon" name="fastforward"></ion-icon>
        </span>
    </div>
    </div>
 </div>
</ion-footer>
```

对应 ts 文件，代码如 CORE0313 所示。

代码 CORE0313　ts 文件

```typescript
import { Component } from '@angular/core';
import { MediaPlugin, MediaObject } from '@ionic-native/media';
@Component({
  templateUrl: 'hello-ionic.html'
})
export class HelloIonicPage {
  constructor(private media: MediaPlugin) {
  }
// 定义音乐对象
file:MediaObject;
playmusic() {
// 设置音乐路径
```

```
    this.file=this.media.create('http://so1.111ttt.com:8282/2017/1/05m/09/298092040183.
m4a?tflag=1496216227&pin=58e48cea3b46ba58686a112cc6df0e73&ip=111.165.165.102#.
mp3');
     this.file.play();
    // 播放音乐
    document.querySelectorAll(".iconCutPlay")[0].setAttribute("hidden","hidden");
    // 隐藏播放按钮
    document.querySelectorAll(".iconCutPause")[0].removeAttribute("hidden");
    // 显示暂停按钮
  }
  pausemusic(){
    this.file.pause();
    // 暂停音乐
    document.querySelectorAll(".iconCutPlay")[0].removeAttribute("hidden");
    // 显示播放按钮
    document.querySelectorAll(".iconCutPause")[0].setAttribute("hidden","hidden");
    // 隐藏暂停按钮
  }
}
```

通过下面十个步骤的操作,实现图 3.2 所示的"Dancer 时代"音频模块界面效果。

第一步:在项目根目录打开命令窗口,运行命令创建音乐视频相关界面,项目二中已经创建视频分类界面和音乐首界面这里就不再进行创建,如图 3.15 所示。

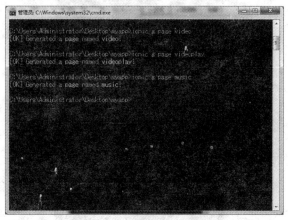

图 3.15　命令窗口创建界面

第二步：打开 Sublime Text 2 软件，进行视频分类界面的制作。该界面由列表图片组成，通过从 ts 中获取 JSON 数据进行列表图片的填充，部分代码如代码 CORE0314 所示。设置样式前效果如图 3.16 所示。

代码 CORE0314　视频分类（videolist.html）

```html
<!-- 头部设置 -->
<ion-header>
  <ion-navbar>
    <ion-title> 分类 </ion-title>
    <!-- 头部中的标题设置 -->
  </ion-navbar>
</ion-header>
<ion-content padding>
  <!-- 最外层 div 框 -->
  <div class="vidlist-div">
    <!-- 遍历循环设置图片的路径并显示在页面上 -->
    <img class="videolist-img" *ngFor="let videolistplay of videolist" src="{{videolistplay.ptimage}}" (click)="video(videolistplay)">
  </div>
</ion-content>
```

图 3.16　视频分类界面设置样式前

设置视频分类界面,需要为图片设置宽高来改变图片的大小。部分代码 CORE0315 如下。设置样式后效果如图 3.17 所示。

```scss
代码 CORE0315    视频分类 SCSS 代码(videolist.scss)
// 最外层 div 的样式
.vidlist-div {
  width: 100%;
  height: 30%;
}
// 图片的样式
.videolist-img {
  margin: 0;
  width: 100%;
  margin-bottom: 2%;
  height: 100%;
  padding: 0;
  border-radius: 5px;
}
```

图 3.17　视频分类界面设置样式后

设置界面进行页面跳转效果,当点击列表图片时,跳转到视频详情页面。部分代码

CORE0316 如下。

代码 CORE0316　　videolist.ts

```typescript
import { Component } from '@angular/core';
import { NavController, NavParams } from 'ionic-angular';
import { VideoPage } from '../video/video';
// 导入视频详情界面
@Component({
  selector: 'page-videolist',
  templateUrl: 'videolist.html',
})
export class VideolistPage {
  //JSON 数据串
  videolist = [
    {
      name: ' 芭蕾舞 ', ptimage: 'assets/images/a3.jpg'
    },
    {
      name: ' 芭蕾舞 ', ptimage: 'assets/images/a4.jpg'
    },
    {
      name: ' 芭蕾舞 ', ptimage: 'assets/images/a5.jpg'
    },
    {
      name: ' 芭蕾舞 ', ptimage: 'assets/images/a6.jpg'
    },
    {
      name: ' 芭蕾舞 ', ptimage: 'assets/images/a7.jpg'
    },
    {
      name: ' 芭蕾舞 ', ptimage: 'assets/images/a8.jpg'
    },
    {
      name: ' 民族舞 ', ptimage: 'assets/images/3.png'
    }
  ];
  constructor(public navCtrl:NavController, public navParams:NavParams) {
  }
```

```
video(videolistplay) {
  // 获取 img 节点
  var img = document.querySelectorAll('img');
  for (var i = 0; i < this.videolist.length; i++) {
    if (this.videolist[i] == videolistplay) {
      console.log(i+":::::::"+img[i].getAttribute("src"))
      var picture=img[i].src;
      // 将图片路径赋值给变量
      this.navCtrl.push(VideoPage,{id:picture});
      // 跳转到视频列表详情界面, 并传值
    }
  }
}
```

第三步：视频详情界面的制作。

视频详情界面，是由顶部图片区域以及图片区域下方的一个列表组成，其中图片区域的图片是由上一个界面传递过来的，列表包括图片、名称和按钮，其代码如 CORE0317 所示。设置样式前效果如图 3.18 所示。

代码 CORE0317　视频详情（video.html）

```html
<!-- 头部设置 -->
<ion-header>
 <ion-navbar>
   <ion-title>video</ion-title>
   <!-- 头部标题设置 -->
 </ion-navbar>
</ion-header>
<ion-content padding>
 <img class="video-img1" src="">
 <!-- 顶部图片 -->
 <ion-item id="video-item" *ngFor="let videolist of video">
    <!-- 下方的列表区域, 循环遍历输出视频图片、名称 -->
   <img class="video-img2" src="{{videolist.ptimage}}">
    <!-- 视频图片 -->
   <p id="video-p">{{videolist.name}}</p>
    <!-- 视频名称 -->
   <button ion-button color="light" outline class="video-button" (click)="videoplay()">播放</button>
```

```
        <!-- 播放按钮 -->
        <video class="video-span" hidden="hidden" src="{{videolist.src}}"></video>
      </ion-item>
  </ion-content>
```

图 3.18 视频详情界面设置样式前

设置视频详情界面,需要为图片设置宽高来改变图片的大小,以及对字体位置的调整,部分代码 CORE0318 如下。设置样式后效果如图 3.19 所示。

代码 CORE0318　视频详情 SCSS 代码（video.scss）

```
// 顶部图片样式
.video-img1 {
  width: 100%;
  height: 40%;
  border-radius: 2%;
}
// 列表的样式
#video-item{
  padding: 0;
}
// 视频图片的美化
.video-img2 {
```

```
    width: 50px;
    height: 50px;
    float: left;
    margin-right: 5%;
}
// 视频名称的美化
#video-p{
    width: 50%;
    float: left;
    font-size: 100%;
    line-height: 200%;
}
// 播放按钮样式
.video-button{
    float: right;
    margin-top: 4%;
    color: #ccc;
    border-color: #ccc;
}
```

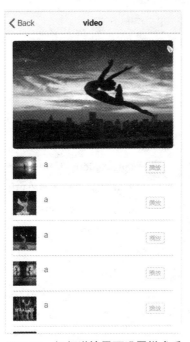

图 3.19　视频详情界面设置样式后

第四步：视频播放插件相关代码如 CORE0319 和 CORE0320 所示。

代码 CORE0319　　命令行安装命令

ionic cordova plugin add https://github.com/moust/cordova-plugin-videoplayer
npm install --save @ionic-native/video-player

代码 CORE0320　　配置及引用代码

```
import { VideoPlayer } from '@ionic-native/video-player';
constructor(private videoPlayer: VideoPlayer) { }
this.videoPlayer.play('file:///android_asset/www/movie.mp4').then(() => {
 console.log('video completed');
}).catch(err => {
 console.log(err);
});
```

第五步：视频播放插件配置。

在 app.module.ts 中进行配置，代码如 CORE0321 所示。

代码 CORE0321　　app.module.ts

```
import { NgModule, ErrorHandler } from '@angular/core';
import { BrowserModule } from '@angular/platform-browser';
import { IonicApp, IonicModule, IonicErrorHandler } from 'ionic-angular';
import { MyApp } from './app.component';
import { TabsPage } from '../pages/tabs/tabs';
import { FriendPage } from '../pages/friend/friend';
import { MinePage } from '../pages/mine/mine';
import { HomePage } from '../pages/home/home';
import { VideoplayPage } from '../pages/videoplay/videoplay';
import { VideoPage } from '../pages/video/video';
import { VideolistPage } from '../pages/videolist/videolist';
import { Core0101Page } from '../pages/CORE0101/CORE0101';
import { RegisterPage } from '../pages/register/register';
import { LoginPage } from '../pages/login/login';
import { AddPage } from '../pages/add/add';
import { StatusBar } from '@ionic-native/status-bar';
import { SplashScreen } from '@ionic-native/splash-screen';
import { VideoPlayer } from '@ionic-native/video-player';
// 需要配置的 ts 路径
@NgModule({
```

```
declarations: [
  MyApp,
  FriendPage,
  MinePage,
  HomePage,
  TabsPage,
  RegisterPage,
  LoginPage,
  AddPage,
  VideolistPage,
  VideoplayPage,
  VideoPage,
  Core0101Page  // 必须引入的参数名
],
imports: [
  BrowserModule,
  IonicModule.forRoot(MyApp,{
    tabsHideOnSubPages: 'true'       // 隐藏全部子页面 tabs
  })
],
bootstrap: [IonicApp],
entryComponents: [
  MyApp,
  FriendPage,
  MinePage,
  HomePage,
  TabsPage,
  RegisterPage,
  LoginPage,
  AddPage,
  VideolistPage,
  VideoPage,
  VideoplayPage,
  Core0101Page  // 必须引入的参数名
],
providers: [
  StatusBar,
  SplashScreen,
```

```
    VideoPlayer,
    {provide: ErrorHandler, useClass: IonicErrorHandler}
  ]
})
export class AppModule {}
```

在播放界面进行配置，代码如 CORE0322。

代码 CORE0322　video.ts

```
import { Component } from '@angular/core';
import { NavController, NavParams } from 'ionic-angular';
import { VideoPlayer } from '@ionic-native/video-player';
// 导入视频播放插件
@Component({
  selector: 'page-video',
  templateUrl: 'video.html',
})
export class VideoPage {
  // 视频列表的 JSON 串
  video =[
    {src: "assets/videos/2.mp4", name: ' 海阳光无 ', ptimage: 'assets/images/a1.jpg'},
    {src: "http://vjs.zencdn.net/v/oceans.mp4", name: ' 白血雨舞 ', ptimage: 'assets/images/a2.jpg'},
    {src: "http://vjs.zencdn.net/v/oceans.mp4", name: ' 爱的心痛 ', ptimage: 'assets/images/a13.jpg'}
  ];
  idd:string;
  // 定义一个 string 类型的全局变量
  constructor(public navCtrl: NavController, public navParams: NavParams, private videoPlayer: VideoPlayer) {
    this.idd=navParams.get("id");
    // 获取上一个界面传递的值
  }
  ionViewDidLoad() {
    var img = document.querySelectorAll('.video-img1');
    // 获取顶部图片节点
    img[0].setAttribute("src",this.idd);
```

```
      // 设置图片的 src 路径
   }
}
```

第六步：添加视频播放功能，当点击播放按钮时进行视频的全屏播放，代码如 CORE0323。

代码 CORE0323　　video.ts 代码

```
import { Component } from '@angular/core';
import { NavController, NavParams } from 'ionic-angular';
import { VideoPlayer } from '@ionic-native/video-player';
// 导入视频播放插件
@Component({
  selector: 'page-video',
  templateUrl: 'video.html',
})
export class VideoPage {
videoplay(){
  var span = document.querySelectorAll('video')[0].src;
  // 获取 src 路径
  this.videoPlayer.play(span).then(() => {
    // 播放视频
    console.log('video completed');
  }).catch(err => {
    console.log(err);
  });
 }
}
```

第七步：音乐首界面的制作。

音乐首界面是由顶部的四个由图标和文字组成的并排的块、中间的本地音乐数量和图标、下面的三个由图标和文字组成块以及最下方的编辑音乐的工具图标组成，部分代码如 CORE0324 所示。设置样式前效果如图 3.20 所示。

代码 CORE0324　　音乐首界面（musicsing.html）

```
<!-- 头部 -->
<ion-header>
 <ion-navbar>
  <ion-title>musicsing</ion-title>
```

```html
    <!-- 头部标题 -->
  </ion-navbar>
</ion-header>
<ion-content class="musicsing-content">
  <div class="music-div">
    <div class="row musicsing-row">
      <!-- 顶部 div-->
      <div class="col">
        <!-- 并列的四个 div-->
        <ion-icon class="miusic-icon1" name="heart-outline"></ion-icon><br>
        <!-- 图标 -->
        <span class="video"> 我喜欢 </span>
        <!-- 文字 -->
      </div>
      <div class="col">
        <ion-icon class="miusic-icon1" name="menu"></ion-icon><br>
        <span class="music"> 歌单 </span>
      </div>
    <div class="col">
        <ion-icon class="miusic-icon1" name="download"></ion-icon><br>
        <span class="download"> 下载 </span>
      </div>
      <div class="col">
        <ion-icon class="miusic-icon1" name="time"></ion-icon><br>
        <span class="picture"> 最近 </span>
      </div>
    </div>
    <ion-item class="musicsing-item" (click)="music()">
      <!-- 本地音乐的设置 -->
      <div class="musicsing-div1">
        <ion-icon class="musicsing-icon" name="phone-portrait"></ion-icon>
        <!-- 图标 -->
        <p>
        <span class="musicsing-span1"> 本地音乐 </span>
          <span class="musicsing-span2">157 首   ></span>
          <!-- 音乐数量 -->
          <ion-icon class="musicsing-icon1" name="play"></ion-icon>
```

```html
      </p>
    </div>
  </ion-item>
</div>
<div class="music-sort">
  <div>
    <span class="music-span1 span1"><ion-icon class="music-icon" name="musical-notes">
    </ion-icon></span>
    <span class="music-span"> 我喜欢 </span>
  </div>
  <div>
    <span class="music-span1 span2"><ion-icon class="music-icon" name="radio">
    </ion-icon></span>
    <span class="music-span"> 我喜欢 </span>
  </div>
  <div>
    <span class="music-span1 span3"><ion-icon class="music-icon" name="people"></ion-icon></span>
  </div>
  <div>
    <ion-icon class="popdiv-icon" name="game-controller-b"></ion-icon>
    <span> 我喜欢 </span>
  </div>
  <div>
    <ion-icon class="popdiv-icon" name="game-controller-b"></ion-icon>
    <span> 我喜欢 </span>
  </div>
  <div>
    <span class="music-span"> 我喜欢 </span>
  </div>
</div>
<!-- 底部工具栏模块 -->
<div class="musicsing-div2">
  <div class="musicsing-div2-div">
    <ion-icon class="musicsing-div2-icon1" name="build"></ion-icon>
    <span> 工具 </span>
```

```
        <ion-icon class="musicsing-div2-icon3" name="arrow-down" (click)="pop()"></ion-icon>
      <ion-icon class="musicsing-div2-icon2" name="notifications-outline"></ion-icon>
    </div>
    <div class="popdiv">
      <div>
        <ion-icon class="popdiv-icon" name="game-controller-b"></ion-icon>
        <span> 我喜欢 </span>
<ion-icon class="popdiv-icon" name="game-controller-b"></ion-icon>
        <span> 我喜欢 </span>
      </div>
      <div>
        <ion-icon class="popdiv-icon" name="game-controller-b"></ion-icon>
        <span> 我喜欢 </span>
      </div>
      <div>
        <ion-icon class="popdiv-icon" name="game-controller-b"></ion-icon>
        <span> 我喜欢 </span>
      </div>
    </div>
  </div>
</ion-content>
```

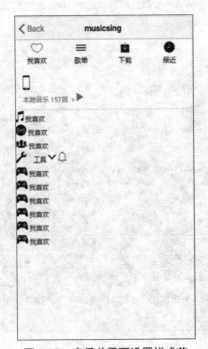

图 3.20　音乐首界面设置样式前

设置音乐首界面，需要设置图标、文字的样式和位置。部分代码如 CORE0325 所示。设置样式后效果如图 3.21 所示。

代码 CORE0325　音乐首界面 SCSS 代码（musicsing.scss）

```scss
// 设置 content
.musicsing-content {
  padding: 0;
}
// 顶部背景的设置
.music-div {
  background: #DFDFE5;
  color: #8e9093;
}
// 顶部并排 div 的样式
.musicsing-row {
  text-align: center;
  border-bottom: 1px solid #F2F2F2;
  padding: 0 -5%;
  height: 15%;
  line-height: 200%;
}
// 图标大小
.miusic-icon1 {
  font-size: 250%;
}
// 本地音乐设置
.musicsing-item {
  padding: 1%;
  background: #DFDFE5;
  color: #8e9093;
}
// 本地音乐中的图标
.musicsing-icon {
  font-size: 200%;
  float: left;
  margin-left: 5%;
}
```

```
// 文字样式
.musicsing-div1 p {
  float: left;
  width: 85%;
  margin-top: 2%;
}
.musicsing-span1 {
  margin-left: 3%;
}
.musicsing-span2 {
  font-size: 70%;
  margin-left: 4%;
}
.musicsing-icon1 {
  float: right;
  margin-right: 5%;
  font-size: 170%;
}
.music-sort {
  text-align: center;
  padding: 0 -5%;
  width: 100%;
  height: 26%;
}
.music-sort div {
  width: 33.3%;
  float: left;
  height: 20%;
}
.music-span1 {
  position: relative;
  height: 30%;
  display: block;
  padding: 30%;
  border-radius: 50%;
  background: #fff;
  width: 30%;
  margin-left: 20%;
```

```css
  margin-top: 10%;
}
.span1 {
  background: orange;
}
.span2 {
  background: #007aff;
}
.span3 {
  background: #f53d3d;
}
.music-icon {
  position: absolute;
  z-index: 2;
  font-size: 200%;
  left: 36%;
  top: 33%;
  color: #FFFFFF;
}
.music-span {
  display: block;
  font-size: 123%;
  margin-top: 10%;
}
// 工具模块样式
.musicsing-div2 {
  position: absolute;
  left: 3%;
  width: 94%;
  border-bottom: 1px solid #dfdfdf;
  border-top: 1px solid #dfdfdf;
  line-height: 40px;
}
.musicsing-div2-div {
  height: 40px;
}
```

```css
.musicsing-div2-div span{
  margin-left: 3%;
}
.musicsing-div2-icon1 {
  margin-left: 5%;
  margin-top: 5px;
  font-size: 150%;
}
.musicsing-div2-icon2 {
  margin-right: 5%;
  font-size: 150%;
  float: right;
  margin-top: 2%;
}
.musicsing-div2-icon3 {
  margin-right: 5%;
  font-size: 150%;
  float: right;
  margin-top: 2%;
}
.popdiv {
  text-align: center;
  display: none;
}
.popdiv div {
  float: left;
  width: 16%;
}
.popdiv-icon {
  display: block;
  line-height: 130%;
}
.popdiv span {
  font-size: 50%;
  line-height: 130%;
  display: block;
}
```

图 3.21　音乐首界面设置样式后

设置音乐界面动画进行页面跳转,当点击最后部分中的向下图标时,弹出更详细的工具。点击中间部分时跳转到音乐播放页面。部分代码如 CORE0326 所示。

代码 CORE0326　音乐首界面跳转代码(musicsing.ts)

```typescript
import { Component } from '@angular/core';
import { NavController, NavParams } from 'ionic-angular';
import { MusicPage } from '../music/music';
// 导入音乐播放界面
@Component({
  selector: 'page-musicsing',
  templateUrl: 'musicsing.html',
})
export class MusicsingPage {

  constructor(public navCtrl: NavController, public navParams: NavParams) {
  }
  music(){
    this.navCtrl.push(MusicPage);
```

```
    // 跳转到音乐播放界面
  }
  istrue:boolean=true;
  pop(){
    if(this.istrue){
      // 正确
      document.querySelectorAll(".popdiv")[0].removeAttribute("hidden");
      // 显示工具
      this.istrue=false;
    }else {
      // 错误
      document.querySelectorAll(".popdiv")[0].setAttribute("hidden","hidden");
      // 隐藏工具
      this.istrue=true;
    }
  }
}
```

第八步:音乐播放界面的制作。

上方内容区域由带有音乐名称和作者的列表组成,下方是音乐播放的一个控制台。部分代码如 CORE0327。设置样式前效果如图 3.22 所示。

代码 CORE0327　音乐播放界面(music.html)

```html
<ion-header>
  <ion-navbar>
    <ion-title>music</ion-title>
  </ion-navbar>
</ion-header>
<ion-content>
<!-- 遍历音乐信息 -->
<div class="playdiv" *ngFor="let ptresourceMP3 of Ptresource">
    <div class="musicdiv" (click)="musicplay(ptresourceMP3)">
      <span class="musicdiv-span1">
      <ion-icon class="musicdiv-icon" name="add"></ion-icon>
      </span>
      <p class="musicname">
        {{ptresourceMP3.musicname}}</p>
```

```html
          <!-- 音乐名称 -->
          <p class="singername">
             {{ptresourceMP3.name}}</p>
          <!-- 作者名字 -->
          <span class="musicdiv-span2">…</span>
       </div>
       <div class="musicDiv" hidden>
         <div></div>
         <img src="{{ptresourceMP3.ptimage}}">
         <!-- 音乐图片 -->
            <p class="musicDiv-p1">
          {{ptresourceMP3.ptmusicname}}</p>
         <!-- 作者+歌曲名称 -->
         <p class="musicDiv-p2">
             <span class="p2-span">
                <ion-icon name="heart-outline"></ion-icon>
             </span>
             <span class="p2-span">
                <ion-icon name="heart-outline"></ion-icon>
             </span>
             <span class="p2-span">
                <ion-icon name="heart-outline"></ion-icon>
             </span>
             <span class="p2-span2">…</span>
         </p>
       </div>
       <audio class="aud" src="{{ptresourceMP3.ptmusic}}"></audio>
    </div>
</ion-content>
<ion-footer>
   <div class="audio-Cut">
      <div class="Cut-div1">
         <img class="musicimage" src="assets/img/yin.jpg">
         <!-- 底部左侧图片 -->
      </div>
      <div class="Cut-div2">
         <div class="jindutiao"><span class="yuandian"></span></div>
         <div class="Cut-div3">
```

```
            <p class="footmusicname"> 成都 </p>
            <!-- 音乐名称 -->
            <p class="footsingername"> 赵雷 </p>
            <!-- 作者名称 -->
        </div>
        <div class="startmodel">
                <span (click)="playmusic()" class="onload">
                    <ion-icon class="iconCutPlay" name="play"></ion-icon>
                    <!-- 播放图标 -->
                    <ion-icon class="iconCutPause" hidden name="pause"></ion-icon>
                    <!-- 暂停图标 -->
                </span>
                <span (click)="downmusic()" class="upmusic">
                    <ion-icon class="upmusic-icon" name="fastforward"></ion-icon>
                    <!-- 下一首图标 -->
                </span>
        </div>
    </div>
 </div>
</ion-footer>
```

图 3.22 音乐播放界面设置样式前

设置音乐播放界面样式，需要设置图片的大小和字体的位置，以及下方播放模块中进度条和图片的位置。部分代码 CORE0328 如下。设置样式后效果如图 3.23 所示。

代码 CORE0328　音乐播放界面 SCSS 代码（music.scss）

```scss
// 最外层 div 设置
.playdiv {
  width: 100%;
  position: relative;
  float: left;
}
// 音乐列表
.musicdiv {
  width: 100%;
  position: relative;
  float: left;
}
// 列表左侧"+"图标的样式
.musicdiv-span1 {
  display: block;
  float: left;
  width: 14px;
  height: 14px;
  border: 1px solid #686868;
  text-align: center;
  line-height: 14px;
  margin-top: 17px;
  margin-left: 3%;
}
// 图标设置
.musicdiv-icon {
  font-weight: bold;
  font-size: 14px;
  color: #686868;
}
// 音乐名称设置
.musicname {
  font-weight: 500;
  color: #3f4959;
```

```css
    width: 80%;
    margin-bottom: 0;
    margin-left: 10%;
    font-size: 16px;
    height: 25px;
    line-height: 30px;
}
// 作者名称设置
.singername {
    color: #7b8f9f;
    width: 80%;
    margin-bottom: 0;
    margin-left: 10%;
    font-size: 14px;
    height: 25px;
    line-height: 20px;
}
// 音乐播放时显示的样式
.musicdiv-span2 {
    width: 6%;
    height: 18px;
    display: block;
    position: absolute;
    right: 10%;
    top: 15px;
    line-height: 9px;
    text-align: center;
    border-radius: 4px;
    border: 1px solid #7b8f9f;
    color: #7b8f9f;
}
// 播放时显示的 div 样式
.musicDiv {
    height: 60px;
    width: 100%;
    position: relative;
    float: left;
}
```

```css
.musicDiv div {
  height: 60px;
  width: 100%;
  position: absolute;
  background: #f8f8f8;
  z-index: -1
}
// 图片设置
.musicDiv img {
  height: 60px;
  width: 60px;
  float: left;
  opacity: 1;
}
.musicDiv-p1 {
  font-weight: 500;
  color: #3f4959;
  width: 70%;
  margin-bottom: 0;
  float: left;
  margin-left: 5%;
  height: 30px;
  line-height: 30px;
  font-size: 16px;
}
.musicDiv-p2 {
  width: 79%;
  height: 30px;
  line-height: 30%;
  text-align: center;
  line-height: 30px;
}
.musicDiv-p2 .p2-span {
  width: 20%;
  display: block;
  float: left;
}
```

```css
.musicDiv-p2 i {
  font-size: 25px;
  color: #7b8f9f;
}
.p2-span2 {
  position: absolute;
  right: 10%;
  top: 35px;
  width: 6%;
  height: 18px;
  display: block;
  line-height: 9px;
  text-align: center;
  border-radius: 4px;
  border: 1px solid #7b8f9f;
  color: #7b8f9f;
}
.audio-Cut {
  width: 100%;
  height: 50px;
  background: #6cb3f3;
}
.Cut-div1 {
  width: 20%;
  position: relative;
  float: left
}
// 底部左侧图标的设置
.Cut-div1 img {
  position: absolute;
  left: 5%;
  top: -30%;
  height: 50px;
  width: 50px;
  border-radius: 50%;
}
```

```css
.Cut-div2 {
  width: 83%;
  position: relative;
  float: right;
  margin-top: 5px;
}
.jindutiao {
  height: 2px;
  width: 95%;
  background: #5795cf;
  position: absolute;
  top: 6px;
  left: 2%;
}
.yuandian {
  display: block;
  height: 4px;
  width: 4px;
  padding: 4px;
  border: 1px solid #fbef7e;
  border-radius: 50%;
  background: #ffc44b;
  position: absolute;
  top: -4px;
  color: #f8f8f8;
}
.Cut-div3 {
  position: absolute;
  left: 3%;
  width: 45%;
  top: 13px;
}
// 歌曲名称设置
.footmusicname {
  font-weight: 500;
  color: #ffffff;
  width: 80%;
  margin-bottom: 0;
```

```
    font-size: 14px;
    height: 16px;
    line-height: 16px;
}
// 作者名字设置
.footsingername {
    color: #c4e0fb;
    margin-bottom: 0;
    font-size: 12px;
    height: 14px;
    line-height: 14px;
}
.startmodel {
    margin-top: 12px;
    color: white;
    float: right;
    position: absolute;
    right: 0;
}
// 播放图标设置
.iconCutPlay {
    color: #ffcd74;
    width: 10px;
    float: left;
    margin-left: -111%;
}
// 暂停图标设置
.iconCutPause {
    color: #ffcd74;
    width: 10px;
    float: left;
    margin-left: -111%;
}
// 下一首图标的设置
.upmusic {
    float: right;
    margin-right: 33%;
}
```

```css
.upmusic-icon {
  color: #ffcd74;
}
```

图 3.23　音乐播放界面设置样式后

第九步：使用音乐播放插件安装并配置，详见技能点 4。

第十步：设置音乐播放效果，当点击列表时，播放音乐，点击下方图标暂停。部分代码如 CORE0329 所示。

代码 CORE0329　音乐播放功能代码（music.ts）

```typescript
import { Component } from '@angular/core';
import { NavController, NavParams } from 'ionic-angular';
import { MediaPlugin, MediaObject } from '@ionic-native/media';
// 导入音乐播放插件
@Component({
  selector: 'page-music',
  templateUrl: 'music.html',
})
export class MusicPage {
```

```
// 音乐信息 json
Ptresource=[
 {
  ptimage: 'assets/img/01.jpg',
  name: ' 赵雷 ',
  musicname: ' 成都 ',
  ptmusicname: ' 赵雷 - 成都 (Live)',
  ptmusic:  'http://so1.111ttt.com:8282/2017/1/05m/09/298092039442.m4a?t-flag=1495810306&pin=048434d16088eba08c67a8674b385d85&ip=61.181.142.83#.mp3',
  selected: true
 },
 {
  ptimage: 'assets/img/02.jpg',
  name: ' 马頔 ',
  musicname: ' 南山南 ',
  ptmusicname: ' 马頔 - 南山南 ',
  ptmusic:  'http://so1.111ttt.com:8282/2017/1/05m/09/298092039442.m4a?t-flag=1495810306&pin=048434d16088eba08c67a8674b385d85&ip=61.181.142.83#.mp3',
  selected: true
 }
];
file:MediaObject;
// 定义一个音乐对象
arr:any;
// 定义一个变量
 constructor(public navCtrl: NavController, public navParams: NavParams,private media: MediaPlugin) {
 }
 ionViewDidLoad() {
  console.log('ionViewDidLoad Music');
 }
 index:any;
// 定义变量
 path:string;
// 定义变量
 musicplay(ptresourceMP3){
   this.arr=[];
```

```
// 赋值
var aud=document.querySelectorAll("audio");
// 取到 audio 节点
for(var i=0;i<aud.length;i++){
  this.arr[i]=true;
}
for (var i = 0; i < this.Ptresource.length; i++) {
  if (this.Ptresource[i] == ptresourceMP3) {
    this.path=aud[i].getAttribute("src");
    // 给 path 赋值
    if(this.index==undefined){
      document.querySelectorAll(".musicdiv")[i].setAttribute("hidden","hidden");
      // 设置隐藏
      document.querySelectorAll(".musicDiv")[i].removeAttribute("hidden");
      // 取消隐藏
      document.querySelectorAll(".footmusicname")[0].innerHTML=document.query SelectorAll(".musicname")[i].innerHTML;
      document.querySelectorAll(".footsingername")[0].innerHTML=document.querySelectorAll(".singername")[i].innerHTML;
    }else {
      this.file.pause();
      document.querySelectorAll(".musicdiv")[this.index].removeAttribute("hidden");
      document.querySelectorAll(".musicDiv")[this.index].setAttribute("hidden","hidden");
      document.querySelectorAll(".musicdiv")[i].setAttribute("hidden","hidden");
      document.querySelectorAll(".musicDiv")[i].removeAttribute("hidden");
      document.querySelectorAll(".footmusicname")[0].innerHTML=document.querySelectorAll(".musicname")[i].innerHTML;
      document.querySelectorAll(".footsingername")[0].innerHTML=document.querySelectorAll(".singername")[i].innerHTML;
    }
    this.index=i;
    this.file=this.media.create("http://so1.111ttt.com:8282/2017/1/05m/09/298092039442.m4a?tflag=1495810306&pin=048434d16088eba08c67a8674b385d85&ip=61.181.142.83#.mp3");
    // 创建音乐,填入路径
    //let duration = this.file.getDuration();
    if(this.arr[i])
```

```
          {
        this.file.play();
        // 播放音乐
        this.arr[i]=false;
        document.querySelectorAll(".iconCutPlay")[0].setAttribute("hidden","hidden");
        // 播放按钮隐藏
        document.querySelectorAll(".iconCutPause")[0].removeAttribute("hidden");
        // 暂停按钮显示
      }
    }
  }
}
playmusic() {
  if(this.arr[this.index]){
    this.file.play();
    // 播放
    this.arr[this.index]=false;
    document.querySelectorAll(".iconCutPlay")[0].setAttribute("hidden","hidden");
    document.querySelectorAll(".iconCutPause")[0].removeAttribute("hidden");
  }else {
    this.arr[this.index]=true;
    this.file.pause();
    // 暂停
    document.querySelectorAll(".iconCutPlay")[0].removeAttribute("hidden");
    document.querySelectorAll(".iconCutPause")[0].setAttribute("hidden","hidden");
  }
}
// 设置下一首
downmusic(){
  this.file.pause();
  var aud=document.querySelectorAll("audio");
  for(var i=0;i<aud.length;i++){
    this.arr[i]=true;
  }
  if(this.index+1<aud.length){
    this.path=aud[this.index+1].getAttribute("src");
```

```
                document.querySelectorAll(".musicdiv")[this.index].removeAttribute("hidden");
                document.querySelectorAll(".musicDiv")[this.index].setAttribute("hidden","hidden");
                document.querySelectorAll(".musicdiv")[this.index+1].setAttribute("hidden","hidden");
                document.querySelectorAll(".musicDiv")[this.index+1].removeAttribute("hidden");
                this.index=this.index+1;
                this.file  =  this.media.create("http://so1.111ttt.com:8282/2017/1/05m/09/298092039442.m4a?tflag=1495810306&pin=048434d16088eba08c67a8674b385d85&ip=61.181.142.83#.mp3");
                if(this.arr[this.index+1])
                {
                this.file.play();
                this.arr[this.index+1]=false;
                document.querySelectorAll(".iconCutPlay")[0].setAttribute("hidden","hidden");
                document.querySelectorAll(".iconCutPause")[0].removeAttribute("hidden");
                document.querySelectorAll(".footmusicname")[0].innerHTML
                document.querySelectorAll(".musicname")[this.index+1].innerHTML;
                document.querySelectorAll(".footsingername")[0].innerHTML=
                document.querySelectorAll(".singername")[this.index+1].innerHTML;
                }
                }
                }
                }
```

至此,"Dancer 时代"音频模块界面及功能基本完成。

本项目通过"Dancer 时代"音频模块的实现,对视频、音频播放功能的实现流程有所了解,掌握 Ionic 弹出框事件及插件的安装及使用,并能够根据所学的 Ionic 弹出框及事件的使用,插件的安装实现 Ionic 更多的功能。

alerts	警报	destructive	破坏性的
dismiss	驳回	swipe	重击
pause	暂停	versions	版本

native　　本机　　　　　　　vibration　　振动
camera　　照相机

一、选择题

1. 以下哪个不是基本的手势事件（　　）。
   A.tap　　　　B.press　　　　C.tab　　　　D.pinch
2. 滑动列表是在 ion-list 标签内添加（　　）标签来使用。
   A.ion-item-sliding　B.<ion-item-options>　C.<ion-item>　D.</ion-list-header>
3. 在视图控制器组件中，确保视图完全呈现的方法是（　　）。
   A.onDidDismiss()　B.isFirst()　　C.hasNavbar()　D.willUnload
4. cordova plugin ls 命令是（　　）。
   A. 删除已安装的插件　　　　　B. 查看所有已经安装的插件
   C. 更新已安装的插件　　　　　D. 安装所需插件
5. 以下哪个不是 Ionic 支持的平台组件方法（　　）。
   A.height()　　B.platforms()　　C.url()　　D.hasNavbar()

二、填空题

1. 基本列表默认情况下，列表之间有分隔线，若要隐藏列表项之间的分隔线，只需在 <ion-list> 加 _____ 属性。
2. Ionic Native 是 Cordova / PhoneGap 插件的一个 _____ 包装器，可以让 Ionic 移动应用程序轻松添加所需的任何本机功能。
3. 如果要安装 Camera 插件，则需要运行 ionic cordova plugin add cordova-plugin-camera 命令，然后使用 _____ 命令安装 Camera 插件。
4. 使用 Camera 插件时，选择返回的图像文件的编码的属性是 _____。
5. 音乐播放暂停播放音频文件的方法是 _____。

三、上机题

使用音乐播放插件实现音乐播放效果。要求：
定义一个音乐列表，当点击列表时进行音乐的播放，效果如下图。

项目三 "Dancer 时代"音频模块的实现 137

项目四 "Dancer 时代"上传模块的实现

通过"Dancer 时代"上传模块的实现,了解选择设备本地文件的实现流程,学习选择文件功能所需插件的相关知识,掌握文件选择插件的使用及 Ionic 的相关选择器,具有独立实现文件选择、图片预览功能的能力。在任务实现过程中:
- 了解选择设备本地文件的实现流程。
- 学习 Ionic 下拉刷新的相关知识。
- 掌握 Ionic 选择器的实际应用。
- 具有独立实现文件选择、图片预览功能的能力。

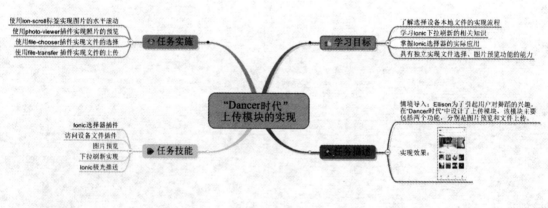

【情境导入】

Ellison 为了引起用户对舞蹈的兴趣,在"Dancer 时代"中设计了上传模块,该模块主要包括两个功能,分别是图片预览和文件上传。用户通过观看舞蹈相关的图片感受舞蹈的魅力,了解舞蹈背后的文化。Ellison 设计的文件上传功能,可以使用户上传自己认为有意义的图片给他人欣赏,用于用户之间的交流。本项目主要通过实现"Dancer 时代"的图片预览和文件上传功能了解 Ionic 的访问本地文件插件和图片预览。

【功能描述】

本项目将实现"Dancer 时代"上传界面。
- 使用 ion-scroll 标签实现图片的水平滚动。
- 使用 photo-viewer 插件实现照片的预览。
- 使用 file-chooser 插件实现文件的选择。
- 使用 file-transfer 插件实现文件的上传。

【基本框架】

基本框架如图 4.1 所示。通过本项目的学习,能将框架图 4.1 转换成效果图 4.2。

图 4.1 框架图　　　　　　　　　　图 4.2 效果图

技能点 1　Ionic 选择器插件

1　日期选择器

日期选择器通常被用于 APP 用户信息填写界面,可以方便的选择日期和时间等信息。其实现方法具有两种：组件和插件。

（1）日期选择器组件

Ionic 提供的 DateTime 组件可以直接设置日期。点击 <ion-datetime> 从下往上滑出选择器界面，在滑出的选择器界面可以选择年、月、日等信息。DateTime 组件具有多种可用格式，其可用格式如下表 4.1 所示。

表 4.1　DateTime 组件可用格式

格　式	描　述
YYYY/YY	年。举例：2017/17
M/MM/MMM/MMMM	月。举例：1⋯12/01⋯12/Jan/January
D/DD/DDD/DDDD	天。举例：1⋯31/01⋯31/Fri/Friday
H/HH	小时，24 小时。举例：0⋯23/00⋯23
h/hh	小时，12 小时。举例：1⋯12/01⋯12
a	12 小时的时间段，小写。举例：am、pm
A	12 小时的时间段，大写。举例：AM、PM
m/mm	分钟。举例：1⋯59/01⋯59
s/ss	秒。举例：1⋯59/01⋯59
Z	UTC 时区偏移。举例：Z or +HH:mm or -HH:mm

对于有些语言难以正确解析 datetime 字符串或格式化 datetime 值。因此 Ionic 使用 ISO 8601 datetime（其格式为 YYYY-MM-DDTHH：mmZ）作为时间选择器格式，并能通过 displayFormat 输入属性设置指定值。ISO 8601 datetime 格式如下表 4.2 所示。

表 4.2　ISO 8601 datetime 格式

格　式	描　述
YYYY	年。举例：1994 年
YYYY-MM	年和月。举例：1994-12
YYYY-MM-DD	完成日期。举例：1994 年 12 月 15 日
YYYY-MM-DDTHH:mm	日期和时间。举例：1994-12-15T13：47
YYYY-MM-DDTHH:mm:ssTZD	UTC 时区。举例：1994-12-15T13：47：20.789Z
YYYY-MM-DDTHH:mm:ssTZD	时区偏移。举例：1994-12-15T13：47：20.789＋5：00
HH:mm	小时和分钟。举例：13:47
HH:mm:ss	小时、分、秒。举例：十三时 47 分二十秒

使用 DateTime 组件效果如图 4.3 所示。

项目四 "Dancer时代"上传模块的实现

图 4.3 日期选择器

为了实现图 4.3 的效果，新建 CORE0401.html，代码如 CORE0401 所示。

代码 CORE0401　日期选择

```
<ion-header>
  <ion-navbar>
    <ion-title>CORE0401</ion-title>
  </ion-navbar>
</ion-header>
<ion-content padding>
  <div> 基本用法 </div>
  <ion-item>
    <ion-label>Date</ion-label>
    <ion-datetime displayFormat="MM/DD/YYYY" [(ngModel)]="myDate">
    </ion-datetime>
  </ion-item>
  <ion-item>
    <ion-label>Date</ion-label>
    <ion-datetime displayFormat="MMM DD, YYYY HH:mm" [(ngModel)]="myDate">
    </ion-datetime>
  </ion-item>
```

```
    <div> 最大最小时间 </div>
    <ion-item>
      <ion-label>Date</ion-label>
        <ion-datetime displayFormat="MMMM YYYY" min="2016" max="2020-10-31" [(ngModel)]="myDate">
        </ion-datetime>
    </ion-item>
  </ion-content>
```

（2）日期选择器插件

日期选择器插件（DatePicker）可以方便的选择对应的时间,在使用之前需要进行安装。在项目目录下打开命令窗口,输入以下命令,安装 DatePicker 插件。

```
ionic cordova plugin ad cordova-plugin-datepicker
npm install –save @ionic-native/date-picker
```

运行命令成功后的效果如图 4.4 所示,安装成功后打开项目中的 plugins 文件夹即可查看如图 4.5 所示内容。

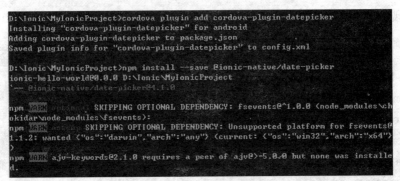

图 4.4　运行命令成功

图 4.5　plugins 文件夹

安装后需要将插件引入项目并进行配置。在 app.module.ts 文件中引入插件,代码如下:

```
import { DatePicker } from '@ionic-native/date-picker';
//...
 providers: [
   StatusBar,
   SplashScreen,
   DatePicker,
   {provide: ErrorHandler, useClass: IonicErrorHandler}
 ]
```

新建界面,创建点击事件,用于调取日期选择器。代码如 CORE0402 所示。

代码 CORE0402   html 文件

```html
<ion-header>
 <ion-navbar>
   <ion-title>CORE0402</ion-title>
 </ion-navbar>
</ion-header>
<ion-content padding>
// 点击事件
<button ion-button (click)="butt()">CORE0302</button>
</ion-content>
```

在对应的 ts 文件中添加实现方法。代码如下所示。

```
import { DatePicker } from '@ionic-native/date-picker';
//...
    constructor(public navCtrl: NavController, public navParams: NavParams,private datePicker: DatePicker) {
  }
  butt(){
   this.datePicker.show({
    date: new Date(),
    mode: 'date',
    androidTheme: this.datePicker.ANDROID_THEMES.THEME_HOLO_DARK
   }).then(
    date => console.log('Got date: ', date),
    err => console.log('Error occurred while getting date: ', err)
   );
  }
```

DatePicker 插件具有多种属性,通过这些属性可以设置不同风格的日期选择器。DatePicker 插件属性如表 4.3 所示。

表 4.3 DatePicker 插件属性

属性	描述
mode	日期选择器的模式值：date | time | datetime
date	选定日期
minDate/maxDate	最小日期默认值 / 最大日期默认值
titleText	标题
okText	确定按钮
cancelText	取消按钮
todayText	当前日期,如果为空,则不显示选择当前日期的选项
nowText	当前时间,如果为空,则不显示选择当前时间的选项
is24Hour	以 24 小时格式显示时间
androidTheme	为选择器选择 Android 主题
allowOldDates	在选择的日期之前显示或隐藏日期
allowFutureDates	显示或隐藏选定日期后的日期
doneButtonLabel	选中按钮
doneButtonColor	选中按钮颜色
cancelButtonLabel	取消按钮
cancelButtonColor	取消按钮颜色
x/y	日期选择器位置,相对位置绝对
minuteInterval	日期选择器分钟部分的选项间隔

2 日历选择器

随着科技水平的提高,纸质版日历已不能满足人们需求,使用 Calendar 插件可以实现日历选择的效果。在使用之前需要进行安装。在项目目录下打开命令提示符,输入以下命令,安装 Calendar 插件。

```
npm install ionic3-calendar --save
```

安装后需要将插件引入项目并进行配置。在 app.module.ts 文件中引入插件,代码如下：

```
import { CalendarModule } from "ionic3-calendar";
//...
imports: [
    IonicModule.forRoot(MyApp),
```

```
    CalendarModule
  ],
```

要想实现日历的效果，需要添加点击事件，用来调取日历选择器，代码如下所示。

```
<button ion-button (click)="openCalendar()">Ionic</button>
```

在对应界面的 ts 文件中添加 openCalendar() 方法。代码如下所示。

```
import { Calendar } from '@ionic-native/calendar';
//...
  constructor(public navCtrl: NavController, public calendarCtrl: CalendarController)
{openCalendar(){
    this.calendarCtrl.openCalendar({
      from:new Date()
    })
    .then( res => { console.log(res) } );
  }
}
```

Calendar 插件具有多种属性，通过这些属性可以定制需求的日历，方便调取日历界面，其属性如表 4.4 所示。

表 4.4 Calendar 插件属性

属　　性	描　　述
from	开始时间
to	结束时间
title	标题
defaultDate	让视图滚动到默认日期
isRadio	是否单选，如果为 false 则选择日期范围
canBackwardsSelected	能否向后选择
closeLabel	取消按钮文字，可以为空
monthTitle	设置月份显示格式
weekdaysTitle	星期显示格式
weekStartDay	周的开始
daysConfig	按天配置
disableWeekdays	自定义 CSS 类，多个用空格分开
cssClass	需要禁用的星期数（0~6，从 0 开始）

使用 Calendar 插件效果如图 4.6 所示。

图 4.6　日历选择器

为了实现图 4.6 的效果，新建界面，代码如 CORE0403 所示。

代码 CORE0403　日历选择

```
<ion-header>
  <ion-navbar>
    <ion-title>Home</ion-title>
  </ion-navbar>
</ion-header>
<ion-content padding>
<button ion-button (click)="btnn()">Ionic</button>
</ion-content>
```

在对应界面的 ts 文件中添加实现方法。代码如下：

```
import { Component } from '@angular/core';
import { NavController } from 'ionic-angular';
import {CalendarController} from "ionic3-calendar/dist";
@Component({
  selector: 'page-home',
  templateUrl: 'home.html'
```

```
})
export class HomePage {
  constructor(public navCtrl: NavController, public calendarCtrl: CalendarController) {
  }
  btnn(){
    this.calendarCtrl.openCalendar({
      from:new Date()
    })
    .then( res => { console.log(res) } );
  }
}
```

3 地区选择器

地区选择器主要作用使用户填写地址时不需要手动输入即可完成填写,其具有操作简单、便于选择等特点。通过单级/多级选择插件即可实现该效果。安装插件代码如下:

```
npm install ion-multi-picker --save
```

安装后需要将插件引入项目并进行配置。在 app.module.ts 文件中引入插件,代码如下所示:

```
import { MultiPickerModule } from 'ion-multi-picker';
//...
imports: [
  BrowserModule,
  IonicModule.forRoot(MyApp),
  MultiPickerModule
],
```

要想实现地区选择的效果,定义一个 JSON 选择器(simpleColumns),代码如下所示。

```
<ion-multi-picker item-content [multiPickerColumns]="simpleColumns"> </ion-multi-picker>
```

在对应界面的 ts 文件中添加选择器的内容。代码如下所示:

```
export class HomePage {
  simpleColumns: any[];
  constructor(public navCtrl: NavController) {
    this.simpleColumns = [
      {
```

```
      name: 'col1',
      options: [{ text: '1', value: '1' },]
    },
    {
      name: 'col2',
      options: [{ text: '1-1', value: '1-1' },]
    },
    {
      name: 'col3',
      options: [{ text: '1-1-1', value: '1-1-1' },]
    }
  ];
 }
}
```

与日期、日历等选择器一样,可以通过多级选择插件实现地区的选择。其属性如表 4.5 所示。

表 4.5 地区选择器属性

属性	描述
multiPickerColumns	配置多选择器列
item-content	添加此属性,以便可以在 ion-item 标记下正确显示此自定义组件
separator	字符串与每列分隔值
cancelText	自定义取消按钮文本
doneText	自定义确定的按钮文本
placeholder	当没有选择选项时,设置占位符文本
text	文本显示在选择器列中
value	文本的关联值
parentVal	指定当前列和上一列之间的依赖关系
disabled	该选项是否可见

地区选择器的实现需要调取对应界面的 ts 文件数据或 JSON 数据,需要有 options 等信息,如表 4.6 所示。

表 4.6 地区选择器属性

选项	描述
options	必需,列中的选项
name	列名

选项	描述
parentCol	当用作依赖选择器时，可以指定父列
alias	列的别名，当使用 parentCol 时，它将找到具有相同名称或别名的列
columnWidth	列的宽度

使用多级选择插件效果如图 4.7 所示。

图 4.7 地区选择器

为了实现图 4.7 的效果，新建界面，代码如 CORE0404 所示。

代码 CORE0404　地区选择

```
<ion-content padding>
  <ion-item>
    <ion-label>Independent Picker</ion-label>
      <ion-multi-picker item-content [multiPickerColumns]="independentColumns" ></ion-multi-picker>
  </ion-item>
  <ion-item>
    <ion-label>Dependent Picker</ion-label>
      <ion-multi-picker id='advanced' item-content [multiPickerColumns]="dependentColumns"></ion-multi-picker>
```

```
    </ion-item>
    <ion-item>
      <ion-label> Simple Picker </ion-label>
      <ion-multi-picker item-content [multiPickerColumns]="simpleColumns">
      </ion-multi-picker>
    </ion-item>
</ion-content>
```

在对应界面的 ts 文件中添加实现方法。代码如下所示。

```
export class HomePage {
 dependentColumns: any[];
 constructor(public navCtrl: NavController) {
  this.dependentColumns = [
    {
      columnWidth: '100px',
      options: [
        { text: '1', value: '1' },
        { text: '2', value: '2' },
        // 代码省略
      ]
    },
    {
      columnWidth: '100px',
      options: [
        { text: '1-1', value: '1-1', parentVal: '1' },
        { text: '1-2', value: '1-2', parentVal: '1' },
        { text: '2-1', value: '2-1', parentVal: '2' },
        { text: '2-2', value: '2-2', parentVal: '2' },
        // 代码省略
      ],
    },
    {
      columnWidth: '100px',
      options: [
        { text: '1-1-1', value: '1-1-1', parentVal: '1-1' },
        { text: '1-1-2', value: '1-1-2', parentVal: '1-1' },
        { text: '2-1-1', value: '2-1-1', parentVal: '2-1' },
        { text: '2-1-2', value: '2-1-2', parentVal: '2-1' },
```

```
       // 代码省略
      ]
    }
  ];
  // 代码省略
  }
}
```

提示 想要了解更多 Ionic 选择器的相关案例,扫描图中二维码,查看更多美观、大方的选择器在不同网站中的应用。快来扫我吧!

技能点 2　访问设备文件插件

随着社交类 APP 的日益增加,发送图片、多媒体等已经成为时代的潮流,为此 Ionic 提供一款插件(访问设备插件),解决访问手机设备中的文件问题。安装插件命令如下。

```
ionic cordova plugin add cordova-plugin-filechooser
npm install --save @ionic-native/file-chooser
```

运行命令成功后的效果如图 4.8 所示,安装成功后打开项目中的 plugins 文件夹即可查看如图 4.9 所示内容。

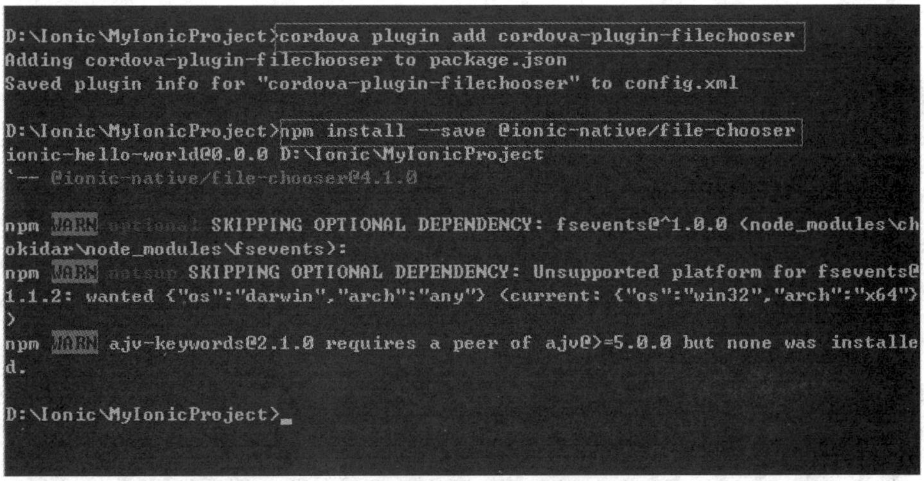

图 4.8　运行命令成功

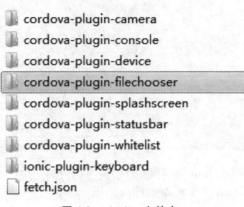

图 4.9　plugins 文件夹

安装后需要将插件引入项目并进行配置。在 app.module.ts 文件中引入插件，代码如下。

```
import { FileChooser } from '@ionic-native/file-chooser';
//...
  providers: [
    StatusBar,
    SplashScreen,
    FileChooser,
    {provide: ErrorHandler, useClass: IonicErrorHandler}
  ]
```

添加点击事件，用于调取选择设备文件界面。代码如下所示。

```
<button ion-button secondary (click)="choosefile()"> 选择文件 </button>
```

在对应的 ts 文件中添加实现方法。代码如下。

```
import { FileChooser } from '@ionic-native/file-chooser';
//...
export class HomePage {
  constructor(public navCtrl: NavController,private fileChooser: FileChooser) {
  }
  butt7(){
this.fileChooser.open()
    .then(uri => console.log(uri))
    .catch(e => console.log(e));
  }
}
```

使用访问设备文件插件效果如图 4.10 所示。

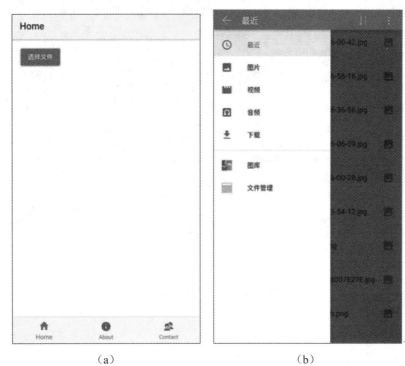

（a）　　　　　　　　　　　　（b）

图 4.10　访问设备文件

为了实现图 4.10 的效果，新建界面，代码如 CORE0405 所示。

代码 CORE0405　html 文件
\<ion-header\>   　\<ion-navbar\>   　　\<ion-title\>Home\</ion-title\>   　\</ion-navbar\>   \</ion-header\>   \<ion-content padding\>   　\<button ion-button (click)="butt7()"\> 选择文件 \</button\>   \</ion-content\>

在对应界面的 ts 文件中添加实现方法。代码如 CORE0406 所示。

代码 CORE0406　ts 文件
import { Component } from '@angular/core';   import { NavController } from 'ionic-angular';   import { FileChooser } from '@ionic-native/file-chooser';   @Component({

```
    selector: 'page-home',
    templateUrl: 'home.html'
})
export class HomePage {
    constructor(public navCtrl: NavController,private fileChooser: FileChooser) {
    }
    butt7(){
      this.fileChooser.open()
        .then(uri => console.log(uri))
        .catch(e => console.log(e));
    }
}
```

技能点 3　图片预览

图片预览是用户点击图片列表中的任意一张图片,点击后图片满屏显示。图片预览功能不仅在手机相册中可以见到,微信查看头像、聊天图片等中也应用广泛。安装命令如下。

```
ionic cordova plugin add com-sarriaroman-photoviewer
npm install --save @ionic-native/photo-viewer
```

安装后需要将插件引入项目并进行配置。在 app.module.ts 文件中引入插件,代码如下。

```
import { PhotoViewer } from '@ionic-native/photo-viewer';
@NgModule({
  providers: [
    StatusBar,
    SplashScreen,
    PhotoViewer
    {provide: ErrorHandler, useClass: IonicErrorHandler}
  ]
})
```

在需要是使用图片预览的界面创建点击事件。代码如下。

```
<img src="image/1.jpg" (click)="pop(image)" >
```

在对应界面的 ts 文件中添加实现方法。代码如下。

```
import { PhotoViewer } from '@ionic-native/photo-viewer';
export class HomePage {
  constructor(private photoViewer: PhotoViewer) {}
pop(){
this.photoViewer.show();
 }
 }
```

其用到的 show(url，title，options) 方法参数如表 4.7 所示。

表 4.7　show 方法选项

事　件	类　型	描　　述
url	string	图片路径
title	string	图片标题
options	any	图片选项，可传递用于隐藏共享按钮的参数 {share:false}

使用图片预览插件效果如图 4.11 所示。

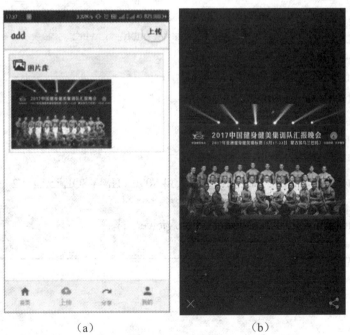

（a）　　　　　　　　　（b）

图 4.11　图片预览

为了实现图 4.11 的效果，新建 home.html，代码如 CORE0407 所示。

代码 CORE0407　图片预览

```html
<div class="add-div1">
  <div class="add-div1-div">
    <ion-icon name="images"></ion-icon>
    图片库
  </div>
  <div class="add-div1-div1">
    <ion-scroll class="add-scroll" scrollX="true">
      <div>
        <img  src="http://119.29.82.34:8090/FHMYSQL/images/2.jpg" (click)="pop(imge)" class="image-list-thumb">
      </div>
    </ion-scroll>
  </div>
</div>
// 省略部分代码
```

在对应界面的 ts 文件中添加实现方法。代码如下。

```typescript
import { Component } from '@angular/core';
import { PhotoViewer } from '@ionic-native/photo-viewer';
@Component({
  selector: 'page-home',
  templateUrl: 'home.html'
})
export class HomePage {
  pop (){
this.photoViewer.show('http://119.29.82.34:8090/FHMYSQL/images/2.jpg.jpg');
  }
  constructor(private photoViewer: PhotoViewer) {}
}
```

技能点 4　下拉刷新实现

1　懒加载

当使用一款软件时，如果界面加载速度过慢，用户会降低使用兴趣，因此，提高界面加载速度在开发中是非常重要的。Ionic 提供的懒加载可以实现该效果，其在用户滑动界面时自动获取新数据，且不会影响原数据的显示，减少了服务端的资源使用。使用懒加载能够减少程序启动时间和打包后的体积，而且可以方便的使用路由功能。Ionic3 的懒加载有如下新特性：
- 避免在 Module 中重复的 import 类。
- 允许通过字符串 key 在任何想使用的地方获取某一个 Page。
- 使代码更简洁。
- 客户响应度更好，体验更友好。
- 开发过程中编译更快。

实现懒加载步骤如下：

第一步：打开 src/app/app.module.ts 文件，移除 app.module.ts 中 declarations，entryComponents 的 HomePage 具体代码如下。

```
@NgModule({
  declarations: [
    MyApp
  ],
  entryComponents: [
    MyApp
  ],
})
```

第二步：创建一个 home.module.ts，引入创建的 HomePage，并通过 IonicPageModule.forChild(OptionsPage) 在全局中声明这个 Page。从 NgModule 模块中删除 HomePage 组件，使该界面的资源在需要时加载。具体代码如下所示。

```
import { NgModule } from '@angular/core';
import { IonicPageModule } from 'ionic-angular';
import { HomePage } from './home';
@NgModule({
```

```
    declarations: [HomePage],
     imports: [IonicPageModule.forChild(HomePage)],
})
    export class HomePageModule { }
```

第三步：在 home.ts 文件中添加 @IonicPage 装饰器，并导入 import { IonicPage } from 'ionic-angular'，代码如下所示。

```
import { IonicPage } from 'ionic-angular'
@IonicPage()
@Component({
  selector: 'page-home',
  templateUrl: 'home.html'
})
export class HomePage {
// 省略部分代码
}
```

第四步：删除 src/app/app.component.ts 文件中的声明，并在 rootPage 中设置为：'HomePage'。代码为：rootPage:any = 'HomePage';。

2　下拉刷新

下拉刷新是向下拉重新加载、刷新，并覆盖在页面上的组件，适用于各种需要内容更新的界面，通常以动态图标、文字等提示的方式显示。当用户获取最新数据后，下拉刷新效果将随即消失。下拉刷新属性如表 4.8 所示。

表 4.8　下拉刷新属性

属　性	描　述
pullingIcon	下拉时显示的图标
pullingText	下拉时显示的文字
refreshingSpinner	刷新时加载指示器 SVG spinner 的名称
refreshingText	刷新时显示的文字

下拉刷新用到 setTimeout(code,millisec) 方法，参数如表 4.9 所示。

表 4.9　下拉刷新 setTimeout() 参数

参　数	描　述
code	必需。执行的 JavaScript 代码
millisec	必需。在执行代码前需等待的毫秒数

使用下拉刷新效果如图 4.12 所示。

为了实现图 4.12 的效果，新建 newsFeed.html，代码如 CORE0408 所示。

对应 ts 文件，代码如 CORE0409 所示。

（a） （b）

图 4.12 下拉刷新

代码 CORE0408　下拉刷新

```html
<ion-list>
  <ion-item *ngFor="let i of items">{{i}}</ion-item>
</ion-list>
<ion-refresher (ionRefresh)="doRefresh($event)">
  <ion-refresher-content
    pullingIcon="arrow-dropdown"
    pullingText="Pull to refresh"
    refreshingSpinner="circles"
    refreshingText="Refreshing...">
  </ion-refresher-content>
</ion-refresher>
```

代码 CORE0409　ts 文件

```
@Component({...})
export class NewsFeedPage {
items = [1];
// doRefresh () 刷新的时候所执行的方法
  constructor(public navCtrl: NavController) {
  }
  doRefresh(refresher) {
    console.log('Begin async operation', refresher);
    setTimeout(() => {
      this.items = [];
      for (var i = 0; i < 30; i++) {
        this.items.push( this.items.length );
      }
      console.log('Async operation has ended');
      refresher.complete();
    }, 2000);
  }
}
```

提示　学会了使用 Ionic 实现下拉刷新，你或许还不知道它的另外实现方式，扫描下方二维码，你会有意想不到的惊喜！

技能点 5　Ionic 极光推送

在市场上有诸多产品提供消息推送服务，例如：百度云、个推、极光等。Ionic 平台提供了一套 jpush 的集成方案——极光推送（JPush），其是一个面向普通开发者免费的第三方消息推送平台，支持 Android、IOS 等系统。使用极光推送步骤如下。

第一步：到极光官网（https://www.jiguang.cn/app/list）注册账号，新建应用获得 appkey，如图 4.13 所示。

点击完成推送设置，如图 4.14 所示。

填写应用包名，如图 4.15 所示。

第二步：引入极光插件。在项目目录下执行如下命令：cordova plugin add jpush-phonegap-plugin --variable API_KEY=xxxxxx（xxxxxx 为极光开发者服务提供的 AppKey）。

第三步：设置启动推送，代码如 CORE0410 所示。

图 4.13　极光官网

图 4.14　推送设置

图 4.15　应用包名

代码 CORE0410　　html 文件

<button ion-button block (click)="initJPush()"> 启动推送 </button>
　<ion-item>
　　<ion-label floating> 别名 Alias</ion-label>
　　<ion-input type="text" [(ngModel)]="alias"></ion-input>
　</ion-item>
　<button ion-button block (click)="setAlias()" [disabled]="alias "> 设置别名

```html
</button>
<ion-list>
  <ion-item text-wrap *ngFor="let msg of msgList">
    <ion-avatar item-left>
      <img src="assets/user.jpg">
    </ion-avatar>
    <h2> 通知 </h2>
    <p>{{msg.content}}</p>
  </ion-item>
</ion-list>
```

对应 ts 文件，代码如 CORE0411 所示。

代码 CORE0411　ts 文件

```typescript
import { Component } from '@angular/core';
import { NavController } from 'ionic-angular';
declare var window;
@Component({
  selector: 'page-home',
  templateUrl: 'home.html'
})
export class HomePage {
  alias: string = '';
  msgList:Array<any>=[];
  constructor(public navCtrl: NavController) {
  }
  initJPush() {
    // 启动极光推送
    if (window.plugins &&    window.plugins.jPushPlugin) {
      window.plugins.jPushPlugin.init();
      document.addEventListener("jpush.receiveNotification", () => {
        this.msgList.push({content:window.plugins.jPushPlugin.receiveNotification.alert})
      }, false);
      alert(" 注册极光推送 ")
    }
  }
  setAlias() {
    // 设置 Alias
```

```
if (this.alias && this.alias.trim() != "") {
  window.plugins.jPushPlugin.setAlias(this.alias);
  alert(" 设置别名 :"+ this.alias);
}else alert('Alias 不能为空 ')
}
}
```

第四步：到极光官网（https://www.jiguang.cn/app/list）发送通知，如图 4.16 所示。

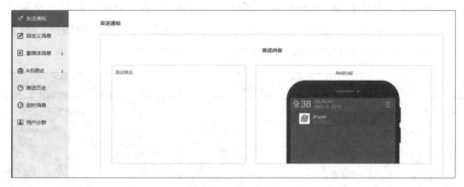

图 4.16　发送通知

测试手机端是否可以收到通知，效果如图 4.17 所示。

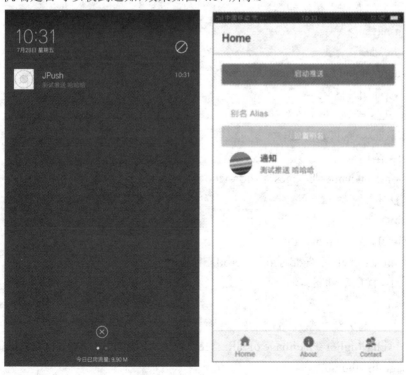

（a）　　　　　　　　　　（b）

图 4.17　测试

通过下面六个步骤的操作,实现图 4.2 所示的"Dancer 时代"上传模块界面及所对应的功能。

第一步:切换到项目目录下,在命令窗口创建上传界面并配置(由于项目二中创建选项卡时已经创建并配置好,这里就不再进行创建)。

第二步:进行上传界面的制作。

界面是由头部左侧的按钮、下方的两个图片展示框组成,第一个展示框是水平滑动进行查看,这样做效果美观并且节省空间,第二个展示框可以直观的展示所有的图片。部分代码如 CORE0412 所示。设置样式前效果如图 4.18 所示。

```
代码 CORE0412   上传页面(add.html)
<!-- 头部 -->
<ion-header>
  <ion-navbar>
    <ion-title>add</ion-title>
    <button class="add-button" ion-button color="light" round (click)="doConfirm()">
    上传 </button>
    <!-- 头部右侧上传按钮 -->
  </ion-navbar>
</ion-header>
<ion-content padding id="add-content">
<!-- 顶部图片区域 -->
<div class="add-div1">
  <div class="add-div1-div">
    <!-- 图标 -->
    <ion-icon name="images"></ion-icon>
    图片库
  </div>
  <div class="add-div1-div1">
    <ion-scroll class="add-scroll" scrollX="true">
      <!-- 横向滚动条 -->
      <div>
        <!-- 遍历获取图片路径 -->
        <img *ngFor="let image of allImages" src="{{image.src}}" (click)="pop(image)" class="image-list-thumb">
```

```
      </div>
    </ion-scroll>
  </div>
</div>
<!-- 底部图片区域 -->
<div class="add-div2">
  <div class="add-div2-div">
    <!-- 图标 -->
    <ion-icon name="images"></ion-icon>
    图片库
  </div>
  <a>
    <div class="col-24" *ngFor="let image of allImages">
      <!-- 遍历循环获取图片路径 -->
      <img src="{{image.src}}" class="image-list-thumb2" (click)="popimg(image)"/>
    </div>
  </a>
</div>
</ion-content>
```

图 4.18 上传界面设置样式前

设置上传界面样式，需要设置上传按钮的样式和位置，设置展示图片的大小和排列方式。部分代码如 CORE0413 所示。设置样式后，如图 4.19 所示。

代码 CORE0413　上传界面 SCSS 代码（add.scss）

```scss
// 上传按钮样式
.add-button {
  margin: 0;
  padding: 3%;
  position: absolute;
  top: 0%;
  right: 1%;
}
//ion-content 的样式
#add-content .scroll-content {
  padding: 0px;
}
// 顶部 div 样式
.add-div1 {
  width: 94%;
  height: 210px;
  border: 1px solid #ccc;
  margin: 3%;
}
.add-div1-div {
  padding: 8px;
  background-color: #f5f5f5;
  color: #222;
  font-weight: 500;
  border-bottom: 1px solid #ccc;
}
.add-div1-div1 {
  height: 80%;
}
// 滚动条样式
.add-scroll {
  width: 100%;
  height: 100%;
```

```css
  overflow: hidden;
}
.add-scroll div {
  width: 175%;
  padding: 2px;
}
// 顶部图片样式
.image-list-thumb {
  padding: 2px;
  height: 150px;
}
// 底部 div 样式
.add-div2 {
  width: 94%;
  height: 260px;
  border: 1px solid #ccc;
  margin: 3%;
}
.add-div2-div {
  padding: 8px;
  background-color: #f5f5f5;
  color: #222;
  font-weight: 500;
  border-bottom: 1px solid #ccc;
}
// 图片框样式
.col-24 {
  padding: 5px;
  border: 1px solid #eee;
  margin-bottom: 10px;
  margin-right: 1%;
  width: 24%;
  display: block;
  float: left;
}
```

```
// 底部图片样式
.image-list-thumb2 {
  padding: 2px 2px 2px 2px;
  height: 80px;
  width: 98%;
}
```

第三步：设置界面弹出框效果，当点击上传按钮时，弹出选择提示框。部分代码如 CORE0414 所示。效果如图 4.20 所示。

图 4.19　上传界面设置样式后

代码 CORE0414　弹出效果

```
import { Component } from '@angular/core';
import { NavController, NavParams, ActionSheetController } from 'ionic-angular';
@Component({
  selector: 'page-add',
  templateUrl: 'add.html',
})
export class AddPage {
  allImages = [{
```

```
      src: 'http://119.29.82.34:8090/FHMYSQL/musics/Data-lb1.jpg'
    }, {
      src: 'assets/img/001.jpg'
    }, {
      src: 'assets/img/02.jpg'
    }, {
      src: 'assets/img/002.jpg'
    }, {
      src: 'assets/img/03.jpg'
    }, {
      src: 'assets/img/b3.jpg'
    }, {
    src: 'assets/img/1.jpg'
    }, {
      src: 'assets/img/2.jpg'
    }];
    constructor(public navCtrl:NavController, public navParams:NavParams, public actionSheetCtrl:ActionSheetController) {
    }
    doConfirm() {
      alert();
      let actionSheet = this.actionSheetCtrl.create({
        title: '文件上传',
        buttons: [
          {
            text: '选择文件',
            handler: () => {
              alert('选择文件');
            }
          },
          {
            text: '上传',
            handler: () => {
              alert('上传');
            }
          },
          {
```

```
          text: '取消',
          role: 'cancel',
          handler: () => {
            alert('取消');
          }
        }
      ]
    });
    actionSheet.present();
  }
```

图 4.20　弹出效果

第四步：文件选择功能，当出现弹出框时，点击上传，弹出文件夹选择。这里我们需要用到文件夹选择插件进行选取。插件安装及使用详见技能点 2。文件选择功能的实现，代码如 CORE0415 所示。效果如图 4.21 所示。

代码 CORE0415　add.ts 代码

```
import { Component } from '@angular/core';
import { NavController, NavParams } from 'ionic-angular';
import { FileChooser } from '@ionic-native/file-chooser';
```

```typescript
// 导入文件选择插件
@Component({
  selector: 'page-add',
  templateUrl: 'add.html',
})
export class AddPage {
  allImages = [{
    src: 'assets/img/01.jpg'
  }
  ];
  constructor(public navCtrl:NavController, public navParams:NavParams, private fileChooser:FileChooser) {
  }
  doConfirm() {
    alert();
    let actionSheet = this.actionSheetCtrl.create({
      title: ' 文件上传 ',
      buttons: [
        {
          text: ' 选择文件 ',
          handler: () => {
            this.fileChooser.open()
              .then(uri => alert(uri))
              .catch(e => alert(e));
          }
        },
        {
          text: ' 上传 ',
          handler: () => {
            alert(' 上传 ');
          }
        },
        {
          text: ' 取消 ',
          role: 'cancel',
          handler: () => {
            alert(' 取消 ');
```

```
                    }
                }
            ]
        });
        actionSheet.present();
    }
```

图 4.21 文件选择

第五步：当点击图片时可以对该图片进行预览查看，实现该效果需要引用图片查看插件（插件使用见技能点 3）。

第六步：把图片预览、文件选取和文件上传效果代码整理进入 add.ts 中，代码如 CORE0416 所示。

代码 CORE416 上传界面 ts 代码（add.ts）

```
import { Component } from '@angular/core';
import { NavController, NavParams, AlertController, ActionSheetController } from 'ionic-angular';
//ActionSheetController 弹出框导入
import {FileChooser} from '@ionic-native/file-chooser';
// 导入文件选择插件
import {PhotoViewer} from '@ionic-native/photo-viewer';
```

```typescript
// 导入图片预览插件
@Component({
  selector: 'page-add',
  templateUrl: 'add.html',
})
export class AddPage {
  allImages = [{
    src: 'http://123.207.150.36:8090/FHMYSQL/musics/Data-lb1.jpg'
  }, {
    src: 'assets/img/001.jpg'
  }, {
    src: 'assets/img/02.jpg'
  }, {
    src: 'assets/img/002.jpg'
  }, {
    src: 'assets/img/03.jpg'
  }, {
    src: 'assets/img/b3.jpg'
  }, {
    src: 'assets/img/1.jpg'
  }, {
    src: 'assets/img/2.jpg'
  }];
    constructor(public navCtrl:NavController, public navParams:NavParams, public alerCtrl:AlertController, private fileChooser:FileChooser, public actionSheetCtrl:ActionSheetController, private photoViewer:PhotoViewer) {
    }
    doConfirm() {
      let actionSheet = this.actionSheetCtrl.create({
        title: ' 文件上传 ',
        buttons: [
          {
            text: ' 选择文件 ',
            handler: () => {
              this.fileChooser.open()
                .then(uri => alert(uri))
```

```
                .catch(e => alert(e));
            }
        },
        {
            text: '上传',
            handler: () => {
            }
        },
        {
          text: '取消',
            role: 'cancel',
            handler: () => {
            }
        }
      ]
    });
    actionSheet.present();
  }
  pictur:string;
  pop(image) {
    var img = document.querySelectorAll('.image-list-thumb');
    for (var i = 0; i < this.allImages.length; i++) {
      if (this.allImages[i] == image) {
        console.log(i + ":::::::" + img[i].getAttribute("src"))
this.pictur = img[i].getAttribute("src");
this.photoViewer.show(this.pictur, 'My image title', {share: false});
      }
    }
  }
}
```

至此,"Dancer 时代"上传模块界面及功能基本完成。

本项目通过"Dancer 时代"上传模块的实现,对选择设备本地文件的实现流程具有一定的了解,对文件选择插件的使用及 Ionic 的相关选择器有所认识,能独立实现文件选择、图片预览等功能。

datetime	日期时间	datePicker	期选择器
mode	模式	calendar	日历
options	选项	JPush	极光推送
variable	变量	millisec	毫秒
code	代码		

一、选择题

1.（　　）插件在 Ionic 2 中主要用于支持访问手机设备中的文件功能。
A.cordova-filechooser　　　　　　B.cordova- barcodescanner
C.cordova-geolocation　　　　　　D.cordova-dialogs

2. 图片预览所使用的插件是（　　）。
A.cordova-plugin-image-resizer　　B.cordova-plugin-console
C.cordova-plugin-device　　　　　D.cordova-plugin-dialogs

3. 图像成功下载后,将其缓存在内存中的属性是（　　）。
A.src　　　　B.Alt　　　　C.bounds　　　　D.cache

4. Ionic 使用（　　）进行数据遍历。
A.()　　　　B.(())　　　　C.{}　　　　D.{{}}

5. Ionic 输入框取值的正确写法是（　　）。
A.ngapp　　　B.ngModel　　　C.[(ngModel)]　　　D.)ng--bing

二、填空题

1. 使用 cordova-filechooser 插件来实现该功能主要有四个步骤,分别为:安装插件,_____,插入方法和_____。

2. _____ 组件的功能包括加载可见的图像和懒加载。

3. Ionic3 添加点击事件的代码格式为:_____。

4. 图像尺寸可以通过_____或外部样式表设置图片的相关属性。

5. 安装插件有两种方法:一种是网站下载,另一种是_____。

三、上机题

编写符合以下要求的网页,实现弹出框并选择文件效果。要求:
1.定义一个按钮,当点击按钮时弹出弹出框

2. 点击选择文件实现文件选取效果，效果如下图。

项目五 "Dancer 时代"分享模块的实现

通过"Dancer 时代"分享模块的实现，了解分享功能的实现流程，学习 Ionic 中指纹验证、社交分享及地图定位等插件，掌握分享插件的使用及地图定位的申请步骤，具有独立使用插件实现分享功能的能力。在任务实现过程中：

- 了解分享功能的实现流程。
- 掌握指纹、分享等插件。
- 掌握地图定位的申请步骤及使用。
- 具有独立使用插件实现分享功能的能力。

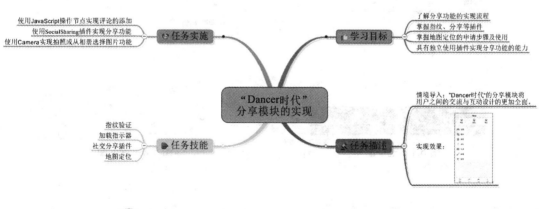

【情境导入】

"Dancer 时代"的分享模块将用户之间的交流与互动设计的更加全面。用户可以将自己感觉幽默的笑话、优美的图片等上传到软件中，好友之间可以互相查看动态、点赞、评论。也可以将动态内容分享到其他软件中,实现了信息共享,解决了用户使用软件的局限性,增强了用户体验。本项目主要通过实现"Dancer 时代"的分享功能了解 Ionic 的社交分享 SocialSharing 插件和地图定位。

【功能描述】

本项目将实现"Dancer 时代"分享模块界面及功能。
- 使用 JavaScript 操作节点实现评论的添加。
- 使用 SocialSharing 插件实现分享功能。
- 使用 camera 实现拍照或从相册选择图片功能。

【基本框架】

基本框架如图 5.1 和图 5.3 所示。通过本项目的学习,能将框架图 5.1 和图 5.3 转换成效果图 5.2 和图 5.4。

图 5.1　框架图 1

图 5.2　效果图 1

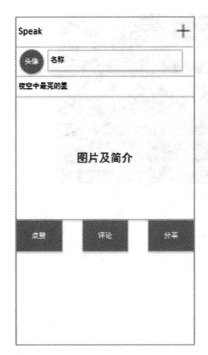

图 5.3　框架图 2　　　　　　　　　图 5.4　效果图 2

技能点 1　指纹验证

手机设备具有多种安全模式,如密码、PIN、指纹验证等。目前,指纹验证被越来越多的用户所使用,用户通过指纹即可进行身份验证,它具有方便、快速、无需记忆即可完成身份验证的特点。在 Ionic 中通过指纹验证插件可以实现这一效果。输入命令安装 Fingerprint Auth 插件。安装 Fingerprint Auth 插件命令如下。

```
ionic cordova plugin add cordova-plugin-android-fingerprint-auth
npm install --save @ionic-native/android-fingerprint-auth
```

运行命令成功后的效果如图 5.5 所示。安装成功后打开项目中的 plugins 文件夹即可查看如图 5.6 所示内容。

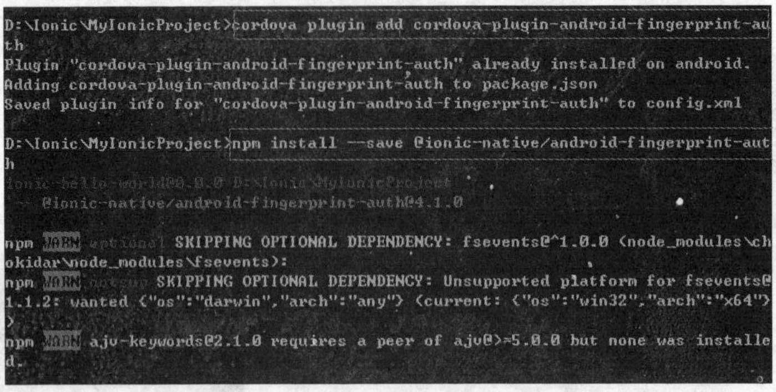

图 5.5 效果图

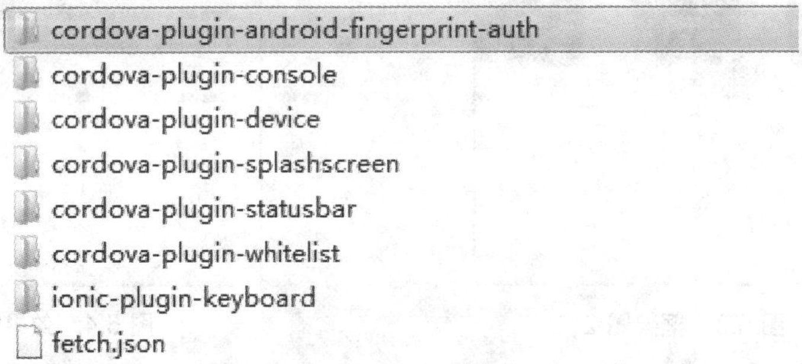

图 5.6 安装完目录

安装后需要将插件引入项目并进行配置。在 app.module.ts 文件中引入插件，代码如下所示。

```
import { AndroidFingerprintAuth } from '@ionic-native/android-fingerprint-auth';
//...
providers: [
    StatusBar,
    SplashScreen,
    AndroidFingerprintAuth,
    {provide: ErrorHandler, useClass: IonicErrorHandler}
]
```

在需要指纹验证的界面创建点击事件，代码如下所示。

```
<button ion-button (click)="butt()"></button>
```

在对应界面的 ts 文件中添加实现方法。代码如下所示。

```
import { AndroidFingerprintAuth } from '@ionic-native/android-fingerprint-auth';
//...
constructor(public navCtrl: NavController,private androidFingerprintAuth: AndroidFingerprintAuth) { }
```

```
butt(){
// 添加想通过事件实现的内容
}
```

指纹验证插件具有多种属性，其属性如表 5.1 所示。

表 5.1 指纹验证属性

属性	描述
clientId	用作 Android Key Store 中密钥的别名
username	用于为加密和别名创建凭证字符串以检索密码
password	用于创建加密的凭据字符串
token	在成功认证时进行解密
disableBackup	设置为 true 作为删除"使用备份"按钮
locale	改变语言
maxAttempts	设备最多为 5 次尝试
userAuthRequired	要求用户使用指纹进行身份验证
dialogTitle	设置指纹认证对话框的标题
dialogMessage	设置指纹认证对话框的消息
dialogHint	在指纹认证对话框中设置指纹图标显示的提示

使用指纹验证插件效果如图 5.7 所示。

图 5.7 指纹验证

为了实现图 5.7 的效果，新建界面，代码如 CORE0501 所示。

代码 CORE0501　html 文件

```html
<ion-header>
 <ion-navbar>
  <ion-title>Home</ion-title>
 </ion-navbar>
</ion-header>
<ion-content padding>
 <h2>欢迎使用 Ionic</h2>
 <button ion-button (click)="butt()"></button>
</ion-content>
```

在对应界面的 ts 文件中添加实现方法。代码如 CORE0502 所示。

代码 CORE0502　ts 文件

```ts
import { Component } from '@angular/core';
import { NavController } from 'ionic-angular';
import { AndroidFingerprintAuth } from '@ionic-native/android-fingerprint-auth';
@Component({
  selector: 'page-home',
  templateUrl: 'home.html'
})
export class HomePage {
  constructor(public navCtrl: NavController,private androidFingerprintAuth:
  AndroidFingerprintAuth) {
  }
  butt(){
    this.androidFingerprintAuth.isAvailable()
    .then((result)=> {
      if(result.isAvailable){
        this.androidFingerprintAuth.encrypt({clientId: 'myAppName', username: 'myUsername', password: 'myPassword' })
        .then(result => {
          if (result.withFingerprint) {
            console.log('Successfully encrypted credentials.');
            console.log('Encrypted credentials: ' + result.token);
          } else if (result.withBackup) {
            console.log('Successfully authenticated with backup password!');
```

```
        } else {console.log('Didn\'t authenticate!')};
      })
      .catch(error => {
          if (error === this.androidFingerprintAuth.ERRORS.FINGERPRINT_CAN-
CELLED) {
           console.log('Fingerprint authentication cancelled');
          } else {console.error(error)};
      });
      } else {
      }
    })
    .catch(error => console.error(error));
  }
```

提示 与其天天在乎自己的成绩和物质利益,不如努力学习、工作。扫描图中二维码,与小和尚一起享受每一次经历的过程,并从中学习成长。

技能点 2 加载指示器

加载指示器是指加载效果覆盖在界面内容上面,以半透明的形式遮挡住界面中的内容,加载结束后关闭指示器,以恢复用户对应用的使用。加载指示器也可在指定时间后自动关闭,默认情况下它会显示为旋转的圆形。加载指示器在页面更改期间也会显示,可通过将 ismissOnPageChange 属性设为 true 来禁用。ionSpinner 提供了许多种旋转加载的动画图标,该图标采用 <ion-spinner> 标签设置,代码如:<ion-spinner icon="spiral"></ion-spinner>。

在实现加载效果中需要用到一些属性来设置加载的效果,具体属性如表 5.2 所示。

表 5.2 加载属性

属性	描述
spinner	设置加载时显示的图标
content	设置加载时显示的内容
showBackdrop	设置是否显示背景。默认为 true
enableBackdropDismiss	设置是否通过点击背景解除指示器。默认为 false
dismissOnPageChange	设置导航到新页面时是否关闭指示器。默认为 false
duration	设置加载时间。默认情况下,它将显示直到 dismiss() 被调用

使用加载指示器效果如图 5.8 所示。

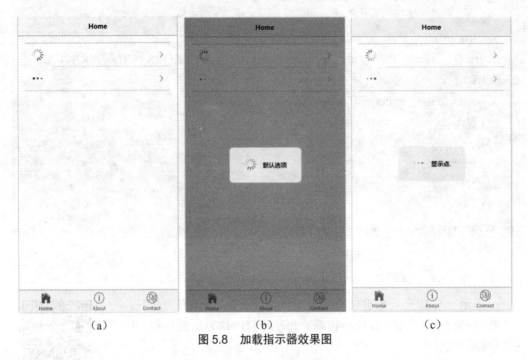

图 5.8 加载指示器效果图

为了实现图 5.8 的效果,新建界面,代码如 CORE0503 所示。

代码 CORE0503　加载效果 (HTML)

```
<ion-header>
 <ion-navbar>
  <ion-title>Home</ion-title>
 </ion-navbar>
</ion-header>
<ion-content padding >
 <ion-list>
  <button ion-item (click)="presentLoadingIos()">
   <ion-spinner item-start name="ios"></ion-spinner>
  </button>
  <button ion-item (click)="presentLoadingDots()">
   <ion-spinner item-start name="dots"></ion-spinner>
  </button>
 </ion-list>
</ion-content>
```

对应界面的 ts 代码如 CORE0504 所示。

代码 CORE0504　加载界面 ts 代码

```typescript
import { Component } from '@angular/core';
import { NavController } from 'ionic-angular';
import { Slides } from 'ionic-angular';
import { ViewChild } from '@angular/core';
import {LoadingController} from "ionic-angular/index";
@Component({
  selector: 'page-home',
  templateUrl: 'home.html'
})
export class HomePage {
  constructor(public navCtrl: NavController,public loadingCtrl: LoadingController) {
  }
  presentLoadingIos() {
    let loading = this.loadingCtrl.create({
      spinner: 'ios',
      content: ' 默认选项 ',
      duration: 7000,
      enableBackdropDismiss:true,
      dismissOnPageChange:true,
    });
    loading.present();
  }
  presentLoadingDots() {
    let loading = this.loadingCtrl.create({
      spinner: 'dots',
      content: ' 显示点 .',
      duration: 3000,
      showBackdrop: false,
    });
    loading.present();
  }
}
```

技能点 3　社交分享插件

在很多社交软件中可以通过网络以各种形式分享文件、图片、网页链接等，分享功能主要是依赖 SocialSharing 插件实现，实现分享功能首先要下载并安装 SocialSharing 插件。在项目目录下打开命令窗口，输入如下命令。

```
ionic cordova plugin add cordova-plugin-x-socialsharing
npm install --save @ionic-native/social-sharing
```

安装后需要将插件引入项目并进行配置。在 app.module.ts 文件中引入插件，代码如下所示。

```
import { SocialSharing } from '@ionic-native/social-sharing';
@NgModule({
  providers: [
    StatusBar,
    SplashScreen,
    SocialSharing,
    {provide: ErrorHandler, useClass: IonicErrorHandler}
  ]
})
```

在需要使用分享插件的界面创建点击事件。代码如下所示。

```
<button ion-button (click)="shareing()"> 分享 </button>
```

在对应界面的 ts 文件中添加实现方法。代码如下所示。

```
shareing(){
  this.socialSharing.canShareViaEmail().then(() => {
  }).catch(() => {
  });
  this.socialSharing.shareViaEmail('Body', 'Subject', ['recipient@example.org']).then(() => {
  }).catch(() => {
  });
}
```

其用到的 share(message,subject,file,url) 方法参数如表 5.3 所示。

表 5.3　share() 方法参数

参　数	类　型	描　述
message	string	分享的消息
subject	string	主题
file	string	文件或图像的 URL
url	string	分享的网址

使用分享插件效果如图 5.9 所示。

图 5.9　分享插件效果图

为了实现图 5.9 的效果，新建界面，代码如 CORE0505 所示。

代码 CORE0505　html 文件

```
<ion-content padding>
    <h1 (click)="shareing()"> 分享 </h1>
</ion-content>
```

在对应界面的 ts 文件中添加实现方法。代码如 CORE0506 所示。

代码 CORE0506　ts 文件

```
import { Component } from '@angular/core';
import { SocialSharing } from '@ionic-native/social-sharing';
@Component({
```

```
    selector: 'page-home',
    templateUrl: 'home.html'
})
export class HomePage {
    constructor(private socialSharing: SocialSharing) {}
    shareing(){
     // 分享插件
     this.socialSharing.share(' 消息 ', 'www.xtgj.com').then(() => {
      // Success!
     }).catch(() => {
      // Error!
     });
    }
}
```

技能点 4 地图定位

在 Ionic3 中使用高德地图需要在高德地图的开发者平台申请使用 JavaScript API 的 Key。然后通过以下步骤进行地图的开发。最终用户通过手机浏览器或 APP 就可以访问地图 JavaScript API 创建出来的应用。在高德地图的开发者平台申请 Key 的步骤如下。

第一步：注册高德开发者，地址为：http://lbs.amap.com/api/javascript-api/summary/，如图 5.10 所示。

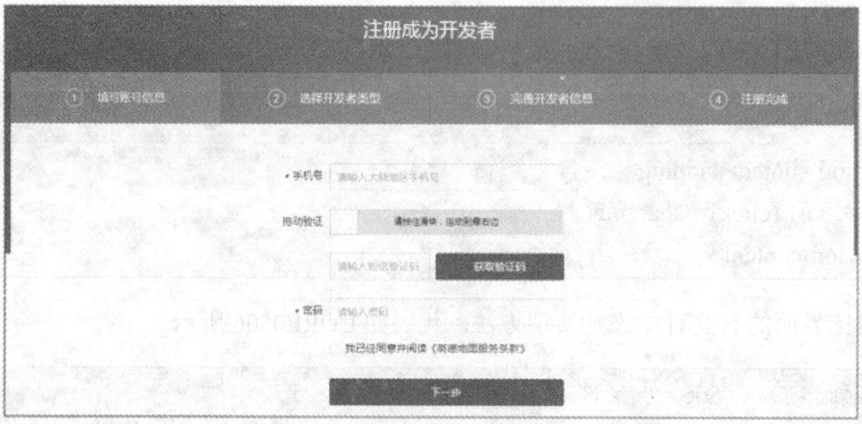

图 5.10 注册高德开发者界面

第二步：去控制台创建应用，效果如图 5.11 所示。

图 5.11 控制台创建应用界面

第三步：获取 Key，界面如图 5.12 所示。

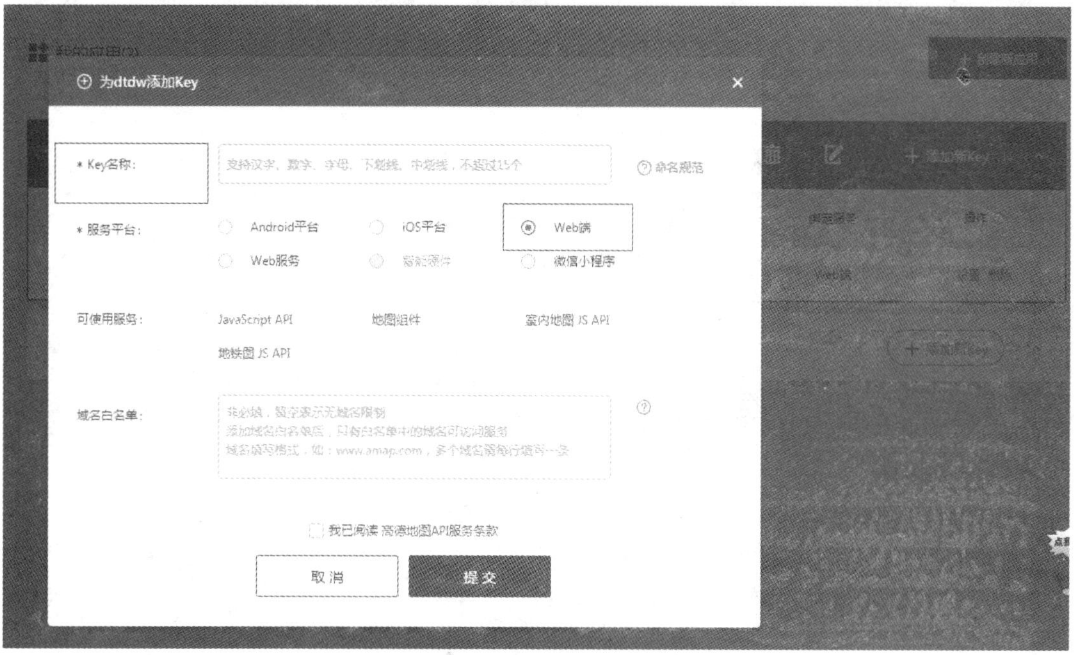

图 5.12 控制台创建应用界面

申请完高德地图的 JavaScript API 的 Key 后，通过以下步骤把高德地图嵌入到 Ionic 项目中，以实现手机 APP 地图定位。

（1）引入高德 JavaScript API，打开 src 目录下的 index.html，在 head 标签中添加以下代码。

```
<script type="text/javascript" src="http://webapi.amap.com/maps?v=1.3&key=您申请的 key 值 "></script>
```

注意：此文件放在 <script src="build/main.js"></script> 的前面。

（2）在 home.html 页面里创建一个 div 容器，并指定 id 标识。

```
<div id="container"></div>
```

（3）在 TypeScript 中接入高德地图类库，声明 AMap 对象。

```
declare const AMap: any;// 声明
```

(4)在 home.ts,添加变量和定义方法

```
public map: any;
// 定义 loadMap() 方法。给地图设定中心点和坐标等属性，center 属性的值可以是
经纬度的二元数组
loadMap() {
this.map = new AMap.Map('container', {
resizeEnable: true,
zoom: 8,
center: [116.39,39.9]
});
}
  ionViewDidLoad() {
// 初始化地图
 let map = new AMap.Map('container', {
   view: new AMap.View2D,
   resizeEnable: true,
   zoom: 11,
   center: [116.397428, 39.90923]
});
// 创建并添加工具条控件 AMap.plugin
AMap.plugin(['AMap.ToolBar'], function () {
  map.addControl(new AMap.ToolBar());
})
//JavaScript API 提供了丰富的覆盖物，比如 Marker 点标记、Polyline 折线、Polygon
多边形、Circle 圆等。以 Marker 为例，我们创建一个最简单的 Marker,并将它添加到地
   图上
let marker = new AMap.Marker({
   position: map.getCenter(),
   draggable: true,
   cursor: 'move'
});
// 也可以在创建完成后通过 setMap 方法执行地图对象
marker.setMap(map);
// 设置点标记的动画效果,此处为弹跳效果
marker.setAnimation('AMAP_ANIMATION_BOUNCE');
}
}
```

使用地图定位效果如图 5.13 所示。

图 5.13 地图定位效果图

为了实现图 5.13 的效果，新建界面，代码如 CORE0507 所示。

代码 CORE0507　html 文件

`<ion-content>`
　`<div id="container" style="height: 500px;" ></div>`
`</ion-content>`

在对应界面的 ts 文件中添加实现方法。代码如 CORE0508 所示。

代码 CORE0508　ts 文件

`import { Component } from '@angular/core';`
`import { NavController } from 'ionic-angular';`
`declare const AMap: any;`// 声明一个对象
`@Component({`
　`templateUrl: 'location.html'`
`})`
`export class LocationPage {`
　`public map: any;`
　`loadMap() {`

```
    this.map = new AMap.Map('container', {
      resizeEnable: true,
      zoom: 8,
      center: [116.39,39.9]
    });
  }
  constructor(public navCtrl: NavController) { }
  ionViewDidLoad() {
    // 初始化地图
    let map = new AMap.Map('container', {
      view: new AMap.View2D,
      resizeEnable: true,
      zoom: 11,
      center: [116.397428, 39.90923]
    });
    AMap.plugin(['AMap.ToolBar'], function () {
      map.addControl(new AMap.ToolBar());
    })
    let marker = new AMap.Marker({
      position: map.getCenter(),
      draggable: true,
      cursor: 'move'
    });
    marker.setMap(map);
    // 设置点标记的动画效果,此处为弹跳效果
    marker.setAnimation('AMAP_ANIMATION_BOUNCE');
  }
}
```

在 src 目录下的 index.html 中引入 JavaScript API。

```
<script type="text/javascript" src="http://webapi.amap.com/maps?v=1.3&key=您申请的 key 值 "></script>
```

提示 当你学会了如何使用地图定位后,你是否还想了解如何启用 / 更改位置服务。扫描图中二维码,让我们学习如何使用 Location Accuracy 插件。

通过下面十二个步骤的操作,实现图 5.2 所示的"Dancer 时代"分享模块效果及功能。

第一步:分享界面的制作。该界面主要是由图片和文字组成,部分代码如 CORE0509 所示。设置样式前效果如图 5.14 所示。

代码 CORE0509　html 代码

```
<ion-header>
 <ion-navbar>
  <!-- 头部设置 -->
  <ion-title>
   About
  </ion-title>
 </ion-navbar>
</ion-header>
<ion-content class="friend-content">
 <!-- 遍历循环填充内容 -->
 <div *ngFor="let item of List">
  <div class="top-div" (click)="speak()">
   <!-- 顶部图标 -->
   <ion-icon class="top-div-icon" name="{{item.icon}}"></ion-icon><br>
   <!-- 名称 -->
   <span>{{item.name}}</span>
  </div>
 </div>
 <!-- 遍历循环填充内容 -->
 <div *ngFor="let items of Lists">
  <ion-item class="bottom-item" (click)="speak()">
   <!-- 底部图标 -->
   <ion-icon name="{{items.icon}}" class="friend-icon"></ion-icon>
   <!-- 名称 -->
   <span>{{items.name}}</span>
  </ion-item>
 </div>
</ion-content>
```

图 5.14 设置样式前效果图

设置分享界面样式,需要设置图标和文字的位置、大小和颜色。部分代码如 CORE0510 所示。设置样式后效果如图 5.15 所示。

```
代码 CORE0510    分享界面样式
//content 样式
.friend-content {
  width: 100%;
  background: #f8f8f8;
}
// 顶部 div 样式
.top-div {
  margin-bottom: 6%;
  text-align: center;
  width: 33.3%;
  background: #FFFFFF;
  height: 10%;
  padding-top: 3%;
  padding-bottom: 3%;
  float: left;
}
```

```css
// 顶部图标样式
.top-div-icon {
  font-size: 250%;
  color: #e08616;
}
// 顶部名称样式
.top-div span {
  font-size: 120%
}
// 底部列表样式
.bottom-item {
  margin: 0;
  padding: 0;
  height: 50px;
  line-height: 50px;
  text-align: center;
}
// 底部图标样式
.friend-icon {
  font-size: 180%;
  line-height: 50px;
  display: block;
  float: left;
  width: 15%;
  color: #a3b736;
}
// 底部名称样式
.bottom-item span {
  font-size: 100%;
  line-height: 50px;
  display: block;
  float: left;
}
```

图 5.15 设置样式后效果图

第二步：创建分享内容界面并进行相关配置。

第三步：进行页面跳转，当点击动态时发生页面跳转。部分代码如 CORE0511 所示。

代码 CORE0511　界面跳转

```
import { Component } from '@angular/core';
import { NavController } from 'ionic-angular';
// 导入 speak
import { SpeakPage } from '../speak/speak';
@Component({
  selector: 'page-about',
  templateUrl: 'friend.html'
})
export class FriendPage {
  // 顶部列表 json
  List=[
    {icon:"logo-xbox",name:" 动态 "},
    {icon:"logo-xbox",name:" 动态 "},
    {icon:"logo-xbox",name:" 动态 "}
  ];
```

```
    // 底部列表 json
    Lists=[
      {icon:"game-controller-b",name:" 游戏 ",cla:"s1"},
      {icon:"american-football",name:" 篮球 ",cla:"s2"},
      {icon:"baseball",name:" 篮球 ",cla:"s3"},
      {icon:"bicycle",name:" 骑行 ",cla:"s4"},
      {icon:"book",name:" 阅读 ",cla:"s5"},
      {icon:"brush",name:" 绘画 ",cla:"s6"},
      {icon:"cart",name:" 购物 ",cla:"s1"}
    ]
    constructor(public navCtrl: NavController) {
    }
    speak(){
      // 跳转到 speak 页面
      this.navCtrl.push(SpeakPage);
    }
  }
```

第四步：分享内容界面制作。

该界面分为三部分：头部、中间用户信息以及底部分享内容。头部是由图标按钮和文字组成，中间用户信息主要是用户的头像与文字介绍，底部是分享内容以及对该内容的按钮操作。部分代码如 CORE0512 所示。设置样式前效果如图 5.16 所示。

代码 CORE0512　设置样式前

```
<ion-header>
  <ion-navbar>
    <ion-title>speak</ion-title>
    <!-- 顶部右侧图标按钮 -->
      <span class="speak-head-span" (click)="add()">
        <ion-icon name="add"></ion-icon></span>
  </ion-navbar>
</ion-header>
<ion-content has-bouncing="true" overflow-scroll="false">
  <!-- 下拉刷新 -->
  <ion-refresher (ionRefresh)="doRefresh($event)">
    <ion-refresher-content></ion-refresher-content>
  </ion-refresher>
  <div class="content">
```

```html
            <!-- 上部div-->
    <div class="content-div">
        <div class="content-top-div">
            <!-- 用户图片 -->
            <img class="photo" src="assets/img/mike.png">
            <!-- 名称 -->
            <h2 class="name-h2"> 夜空中最亮的星 </h2>
            <!-- 相关信息 -->
            <div>
                <a href="#" class="subdued"> 关注：300</a>
                <a href="#" class="subdued"> 粉丝：300</a>
                <p> 天津市，河东区，大王庄 </p>
            </div>
        </div>
        <ul class="list-ul">
          <li>
            <a class="item-a" (click)="jump()">
              <!-- 名称 -->
              <h2 class="title-name"> 夜空中最亮的星 </h2>
              <!-- 图片 -->
              <ion-slides slide-interval="100" does-continue="true" show-pager="false">
                <ion-slide>
                  <img class="full-image photolist" src="assets/img/b1.jpg">
                </ion-slide>
                <ion-slide>
                  <img class="full-image" src="assets/img/b2.jpg">
                </ion-slide>
                <ion-slide>
                  <img class="full-image" src="assets/img/b3.jpg">
                </ion-slide>
              </ion-slides>
              <!-- 内容 -->
              <span class="dm-nowrap-2"> 绕着幸福的公园散步，撷取自在的芬芳；围着快乐的大道锻炼，收获健康的因子；对着挺拔的青山放歌，奏响活力的节拍。全民健身日到了，快来运动吧，让健康永驻身旁。</span>
            </a>
        <div class="content-bottom-div">
            <!-- 点赞 -->
```

项目五 "Dancer 时代"分享模块的实现

```
            <a class="tab-item" (click)="good()">
    <ion-icon name="thumbs-up"></ion-icon>
        点赞 (<span class="number"> 0 </span>)
        </a>
        <!-- 评论 -->
        <a class="tab-item" (click)="jump()">
          <ion-icon name="create"></ion-icon>
         评论 (300)
         </a>
         <!-- 分享 -->
         <a class="tab-item" (click)="show()">
           <ion-icon name="send"></ion-icon>
          分享 (1)
          </a>
        </div>
      </li>
    </ul>
   </div>
  </div>
</ion-content>
```

图 5.16 设置样式前效果图

设置分享内容界面样式,首先设置头部图标按钮的位置、大小和颜色,然后设置中间用户信息的图片大小、圆角和字体大小、颜色、位置,最后设置底部分享内容的图片大小、字体大小、颜色以及下边按钮区域图标和文字大小、位置、颜色。部分代码如 CORE0513 所示。设置样式后效果如图 5.17 所示。

代码 CORE0513　　CSS 代码

```css
// 顶部按钮样式
.speak-head-span {
  float: right;
  display: block;
  width: 28px;
  text-align: center;
  position: absolute;
  right: 0;
  top: 9px;
}
.content {
  overflow: inherit
}
// 中间内容的样式
.content-div {
  box-shadow: 0 1px 3px rgba(0, 0, 0, 0.3);
  margin: 20px 10px;
  border-radius: 2px;
  background-color: #fff;
}
// 用户信息 div 样式
.content-top-div {
  border-color: #ddd;
  background-color: #fff;
  color: #444;
  position: relative;
  z-index: 2;
  display: block;
  margin: -1px;
  padding: 16px;
  border-width: 1px;
  border-style: solid;
```

```
    font-size: 16px;
    padding-left: 72px;
}
// 用户头像样式
.photo {
    position: absolute;
    top: 16px;
    left: 16px;
    max-width: 40px;
    max-height: 40px;
    width: 100%;
    height: 100%;
    border-radius: 50%;
}
// 用户名称样式
.name-h2 {
    margin: 0 0 2px 0;
    font-size: 16px;
    font-weight: normal;
}
// 用户相关信息样式
.content-top-div div {
    color: #666;
    font-size: 14px;
    margin-bottom: 2px;
}
.subdued {
    padding-right: 10px;
    color: #888;
    text-decoration: none;
}
.content-top-div p {
    padding-right: 10px;
    color: #666;
    font-size: 14px;
    margin-bottom: 2px;
    line-height: 26px;
}
```

```css
//ul 样式
.list-ul {
  list-style: none;
}
//ul 下 li 的样式
.list-ul li {
  padding-bottom: 10%;
  border-bottom: 1px solid #F4C63f;
  margin-bottom: 10px;
  margin-left: 0px;
  position: relative;
}
// 发表内容样式
.item-a {
  border-color: #ddd;
  background-color: #fff;
  color: #444;
  position: relative;
  z-index: 2;
  display: block;
  margin: -1px;
  padding: 16px;
  border-width: 1px;
  border-style: solid;
  font-size: 16px;
}
// 发布用户名称
.title-name {
  font-weight: normal;
  font-size: 16px;
  margin-top: 2%;
  margin-bottom: 5%;
}
// 发表的图片
.full-image {
  width: 100%;
  height: 230px;
}
```

```css
// 发表的文字样式
.dm-nowrap-2 {
  font-size: 12px;
  text-overflow: ellipsis;
  overflow: hidden;
  white-space: nowrap;
  width: 96%;
  display: block;
}
// 底部 div 样式
.content-bottom-div {
  border-color: #ddd;
  background-color: #fff;
  color: #444;
  position: relative;
  z-index: 2;
  display: block;
  margin: -1px;
  border-width: 1px;
  border-style: solid;
  font-size: 16px;
  line-height: 49px;
  height: 49px;
}
//div 下按钮的样式
.content-bottom-div a {
  width: 33.3%;
  height: 100%;
  color: inherit;
  text-align: center;
  text-decoration: none;
  display: block;
  float: left;
  font-weight: 400;
  font-size: 10px;
  opacity: 0.7;
}
```

图 5.17 设置样式后效果图

第五步:创建内容详情界面和发布内容界面并配置。

第六步:安装分享插件(详见技能点 3)。

第七步:设置分享内容界面功能,当点击点赞时,后面数字就会加 1,点击评论和中部的介绍区域会发生跳转,跳转到内容详情界面,当点击顶部图标按钮时,会跳转到内容发布界面,最后点击分享时,弹出具有分享方式的弹出框。部分代码如 CORE0514 所示。

```
代码 CORE0514    ts 代码
import { Component } from '@angular/core';
import { NavController, NavParams } from 'ionic-angular';
// 导入分享插件
import { SocialSharing } from '@ionic-native/social-sharing';
// 导入内容详情界面
import { SpeakmessagePage } from '../speakmessage/speakmessage';
// 导入发表内容界面
import { AddspeakPage } from '../addspeak/addspeak';
@Component({
  selector: 'page-speak',
  templateUrl: 'speak.html',
})
export class SpeakPage {
  constructor(public navCtrl: NavController, public navParams: NavParams,private
```

```
socialSharing: SocialSharing) {
}
ionViewDidLoad() {
  console.log('ionViewDidLoad Speak');
}
// 刷新
doRefresh(refresher) {
  console.log('Begin async operation', refresher);

  setTimeout(() => {
    console.log('Async operation has ended');
    refresher.complete();
  }, 2000);
}
jump(){
  // 取到的用户名称
  var name=document.querySelectorAll(".name-h2")[0].innerHTML;
  // 用户头像
  var src=document.querySelectorAll(".photolist")[0].getAttribute("src");
  // 内容图片
  var photo=document.querySelectorAll(".photo")[0].getAttribute("src");
  // 内容文字
  var text=document.querySelectorAll(".dm-nowrap-2")[0].innerHTML;
  // 标题名称
  var titlename=document.querySelectorAll(".title-name")[0].innerHTML;
  this.navCtrl.push(SpeakmessagePage,{ia:name,ib:src,ic:text,id:photo,ie:titlename})
}
good(){
  // 点赞
  var num = document.querySelectorAll('.number')[0];
  var amount=parseInt(num.innerHTML)+1;
  num.innerHTML=" "+amount+" ";
}
show(){
  // 分享插件
  this.socialSharing.share('message', 'subject', 'file', 'url').then(() => {
    // Success!
```

```
     }).catch(() => {
       // Error!
     });
   }
   add(){
     // 跳转到添加发表内容
     this.navCtrl.push(AddspeakPage);
   }
 }
```

第八步:详细内容界面的制作。

该界面由头部的图片展示、中部的人物介绍、底部的评论区域组成。部分代码如 CORE0515 所示。设置样式前效果如图 5.18 所示。

代码 CORE0515　详细内容界面

```html
<ion-header>
  <ion-navbar>
    <!-- 标题 -->
    <ion-title>speakmessage</ion-title>
  </ion-navbar>
</ion-header>
<ion-content>
  <div class="me-con-div">
    <div class="me-div1">
      <!-- 用户名 -->
      <h2 class="me-h2"></h2>
      <!-- 头像 -->
      <img class="full-image" src="assets/img/mike.png">
    </div>
    <div class="me-div2">
      <!-- 头像 -->
      <img src="assets/img/mike.png">
      <!-- 名称 -->
      <h2>Marty</h2>
      <!-- 相关信息 -->
      <p>
        <a href="#" class="subdued"> 关注：300</a>
        <a href="#" class="subdued"> 粉丝：300</a>
```

```html
        </p>
      </div>
      <div class="me-div3">
        <h2>详情介绍:</h2>
        <!-- 内容 -->
        <p class="textp"><!-- 文字省略 --></p>
      </div>
      <h3>   评论:</h3>
      <div>
        <ul class="list-li">
          <li>
            <img src="assets/img/mike.png">
            <span> 匿名网友:这个活动好啊.... 我一定会参加的.... 好奇怪啊放到冯绍峰阿斯顿分分发卡飞机;的看法加 </span>
          </li>
          <li>
            <img src="assets/img/mike.png">
            <span> 匿名网友:这个活动好啊.... 我一定会参加的.... 好奇怪啊放到冯绍峰阿斯顿分分发卡飞机;的看法加 </span>
          </li>
          <li>
            <img src="assets/img/mike.png">
            <span> 匿名网友:这个活动好啊.... 我一定会 dfasdffafdfaf 参加的.... 好奇怪啊放到冯绍峰阿斯顿分分发卡飞机;的看法加 </span>
          </li>
        </ul>
      </div>
    </div>
</ion-content>
<ion-footer class="spme-foot">
  <div>
    <span>
      <ion-icon name="create"></ion-icon>
    </span>
    <!-- 评论内容 -->
    <input [(ngModel)]="content" class="textbox" type="text" placeholder=" 说点什么 ">
  </div>
  <button ion-button (click)="certainly()"> 发送 </button>
</ion-footer>
```

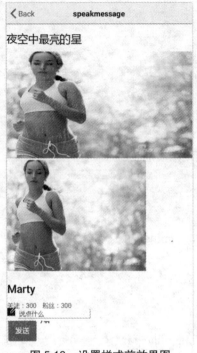

图 5.18　设置样式前效果图

设置详细内容界面样式,首先设置头部图片的大小和位置,然后设置中部内容字体的颜色和大小,之后设置评论区域图片的大小、位置和文字的大小、颜色,最后设置输入框的样式。部分代码如 CORE0516 所示。设置样式后效果如图 5.19 所示。

代码 CORE0516　样式代码

```
// 顶部 div
.me-con-div {
  margin-bottom: 80px;
}
// 用户 div
.me-div1 {
  padding: 16px;
}
// 用户名样式
.me-div1 h2 {
  margin: 10px 0;
  font-size: 16px;
  font-weight: normal;
}
```

```css
// 用户头像样式
.me-div1 img {
  width: 100%;
  height: 230px;
}
// 发表用户 div
.me-div2 {
  border-color: #ddd;
  background-color: #fff;
  color: #444;
  position: relative;
  z-index: 2;
  display: block;
  margin: -1px;
  padding: 16px;
  border-width: 1px;
  padding-left: 72px;
  min-height: 72px;
  border-style: solid;
  font-size: 16px;
}
// 发表用户头像
.me-div2 img {
  position: absolute;
  top: 16px;
  left: 16px;
  max-width: 40px;
  max-height: 40px;
  width: 100%;
  height: 100%;
  border-radius: 50%;
}
// 发表用户名称
.me-div2 h2 {
  margin: 0 0 2px 0;
  font-size: 16px;
  font-weight: normal;
}
```

```css
// 发表用户相关信息
.me-div2 a {
  padding-right: 10px;
  color: #888;
  font-size: 14px;
  text-decoration: none;
}
// 详情介绍 div
.me-div3 {
  border-color: #ddd;
  background-color: #fff;
  color: #444;
  position: relative;
  z-index: 2;
  display: block;
  margin: -1px;
  padding: 16px;
  border-width: 1px;
  border-style: solid;
  font-size: 16px;
}
.me-div3 h2 {
  font-size: 16px;
  font-weight: normal;
  margin-top: 10px;
}
// 内容
.me-div3 p {
  margin-top: 16px;
  color: #666;
  font-size: 14px;
}
// 评论列表
.list-li li {
  padding: 0;
  margin-bottom: .5rem;
  margin-left: .5rem;
  position: relative;
```

```css
    border-bottom: 1px solid lightgray;
}
.list-li img {
    position: absolute;
    top: 0px;
    left: 0px;
    width: 32px;
    height: 32px;
    border-radius: 32px;
}
.list-li span {
    line-height: 32px;
    margin-left: 38px
}
// 评论输入框
.spme-foot {
    border-top: 1px solid #ccc;
}
.spme-foot div {
    width: 80%;
    float: left;
    margin: 7px;
    background: #cccccc;
    border-radius: 3px;
}
.spme-foot span {
    display: block;
    float: left;
    height: 30px;
    line-height: 35px;
    margin: 0 7px;
    font-size: 10px;
}
// 输入内容
.spme-foot input {
    height: 30px;
    background: transparent;
    width: 87%;
```

```
    padding: 0;
    margin: 0;
    border: none;
}
.spme-foot button {
    width: 14%;
    height: 30px;
    margin: 7px 0;
}
```

图 5.19 设置样式后效果图

第九步：设置详细内容界面功能，当在输入框输入文字后，点击发送按钮会在评论的最后面添加该内容。部分代码如 CORE0517 所示。设置详细内容界面效果如图 5.20 所示。

代码 CORE0517　　ts 代码

```
import { Component } from '@angular/core';
import { NavController, NavParams } from 'ionic-angular';
@Component({
  selector: 'page-speakmessage',
  templateUrl: 'speakmessage.html',
})
```

```
export class SpeakmessagePage {
  // 定义变量
  ia:string;
  ib:string;
  ic:string;
  id:string;
  ie:string;
  constructor(public navCtrl: NavController, public navParams: NavParams) {
    // 获取界面传值
    this.ia=navParams.get("ia");
    this.ib=navParams.get("ib");
    this.ic=navParams.get("ic");
    this.id=navParams.get("id");
    this.ie=navParams.get("ie");
  }
  ionViewDidLoad() {
    // 设置用户名称
    document.querySelectorAll(".me-h2")[0].innerHTML=this.ie;
    // 设置内容
    document.querySelectorAll(".textp")[0].innerHTML=this.ic;
    // 设置用户头像
    document.querySelectorAll(".full-image")[0].setAttribute("src",this.ib);
  }
  content:string;
  certainly(){
    var tex=this.content;
    // 创建 li
    var li=document.createElement("LI");
    // 创建 img
    var img=document.createElement("IMG");
    // 设置 img 下的 src
    img.setAttribute("src",this.id);
    // 将 img 上树到 li
    li.appendChild(img);
    // 创建 span
    var span=document.createElement("SPAN");
    // 设置内容
    var text1=document.createTextNode(this.ia+" : "+tex);
```

```
    // 将内容填入 span
    span.appendChild(text1);
    //span 上树到 li
    li.appendChild(span);
    var ul=document.querySelectorAll(".list-li")[0];
    // 将 li 上树到 ul
    ul.appendChild(li);
    this.content="";
  }
}
```

图 5.20　设置详细内容界面效果图

第十步：内容发布界面的制作。

界面是由头部的发布按钮、中间的文字输入区域和底部的图片展示区域组成。部分代码如 CORE0518 所示。设置样式前效果如图 5.21 所示。

代码 CORE0518　发布界面

```
<ion-header>
  <ion-navbar>
    <ion-title>addspeak</ion-title>
```

```html
    <!-- 头部按钮 -->
    <button class="addsp-btn" (click)="push()"> 发布 </button>
  </ion-navbar>
</ion-header>
<ion-content>
  <div class="addspeak-div">
   <!-- 发布文字 -->
   <textarea class="textarea"></textarea>
   <div class="divpreview">
    <!-- 图片框 -->
    <img class="imgpreview" hidden="hidden" src="">
    <span class="videopreview"></span>
   </div>
   <div class="choosediv" (click)="show()">
    <!-- 图标 -->
    <ion-icon id="addspeak-icon" name="camera"></ion-icon><br>
    <span> 照片 / 视频 </span>
   </div>
  </div>
</ion-content>
```

图 5.21　设置样式前效果图

设置内容发布界面样式，首先设置头部发布按钮的大小和位置，然后设置中间部分文字输入框的大小，最后设置底部展示框的大小、位置和点击按钮的样式。部分代码如 CORE0519 所示。设置样式后效果如图 5.22 所示。

代码 CORE0519　　LSS 代码
```
// 头部按钮样式
.addsp-btn{
  position: absolute;
  right: 0;
  width: 40px;
  height: 30px;
  top: 10px;
  border-radius: 10px;
  background: transparent;
}
// 最外层 div
.addspeak-div {
  width: 100%;
  padding-bottom: 20px;
}
// 文本框
.textarea {
  width: 100%;
  height: 150px;
  border: none;
  outline: none;
  padding: 10px;
}
// 图片框
.imgpreview {
  width: 100px;
  height: 100px;
  float: left;
  margin-left: 10px;
  margin-top: 10px;
  border: none;
}
```

```css
// 按钮
.choosediv {
  float: left;
  margin-left: 10px;
  margin-top: 10px;
  width: 100px;
  height: 100px;
  background: #f8f8f8;
  text-align: center;
  color: #cfcfcf;
}
// 图标
#addspeak-icon {
  line-height: 60px;
  font-size: 50px;
}
.choosediv span {
  display: block;
}
```

图 5.22　设置样式后效果图

第十一步：安装拍照插件，详见项目三。

第十二步：设置内容发布界面功能，当点击图标按钮后，弹出提示框，之后点击相册或拍照进入相应的功能。部分代码如 CORE0520 所示。内容发布界面效果如图 5.23 所示。

代码 CORE0520　　ts 代码

```ts
import { Component } from '@angular/core';
import { NavController, NavParams, ActionSheetController } from 'ionic-angular';
// 导入相机/相册插件
import { Camera, CameraOptions } from '@ionic-native/camera';
// 导入 speak 页面
import { SpeakPage } from '../speak/speak';
@Component({
  selector: 'page-addspeak',
  templateUrl: 'addspeak.html',
})
export class AddspeakPage {
  constructor(public navCtrl: NavController, public navParams: NavParams, public actionSheetCtrl: ActionSheetController, private camera: Camera) {
  }
  ionViewDidLoad() {
    console.log('ionViewDidLoad Addspeak');
  }
  // 弹出框
  show(){
    let actionSheet = this.actionSheetCtrl.create({
      buttons: [
        {
          text: ' 相册 ',
          handler: () => {
            const options: CameraOptions = {
              quality: 100,
              destinationType: this.camera.DestinationType.NATIVE_URI,
              saveToPhotoAlbum: true,
              sourceType: this.camera.PictureSourceType.PHOTOLIBRARY
            };
```

```
        // 调取摄像头
        this.camera.getPicture(options).then((imageData) => {
          var ss=document.querySelectorAll(".imgpreview");
            ss[0].removeAttribute("hidden")
            var sss=imageData;
            ss[0].setAttribute("src",sss);
            alert(imageData);
          }, (err) => {
              });
          }
        },
        {
          text: ' 拍照 ',
          handler: () => {
            const options: CameraOptions = {
              quality: 100,
              destinationType: this.camera.DestinationType.NATIVE_URI,
              saveToPhotoAlbum: true,
              sourceType: this.camera.PictureSourceType.CAMERA
            };
        this.camera.getPicture(options).then((imageData) => {
              var ss=document.querySelectorAll(".imgpreview");
              ss[0].removeAttribute("hidden")
              var sss=imageData;
              ss[0].setAttribute("src",sss);
              alert(imageData);
            }, (err) => {
              });
            }
        },
        {
            text: ' 取消 ',
          role: 'cancel',
          handler: () => {
            console.log('Cancel clicked');
```

```
        }
      }
    ]
  });
  actionSheet.present();
}
push(){
  this.navCtrl.push(SpeakPage);
 }
}
```

图 5.23 内容发布界面效果图

至此,"Dancer 时代"分享模块界面及功能基本完成。

本项目通过"Dancer 时代"分享模块界面及功能的学习,对分享功能的实现流程具有一定的了解,对 Ionic 中指纹验证、社交分享及地图定位等插件有所认识,掌握了分享插件的使用及地图定位的申请步骤。

fingerprint Auth	指纹认证	spinner	微调
locale	现场	SocialSharing	社会共享
duration	期间	draggable	拖动
message	消息	declare	声明

一、选择题

1．ionSpinner 提供了许多种旋转加载的动画图标。当界面加载呈现给用户相应的加载图标。该图标采用的是 SVG 格式，代码为（　　）。

A.<ion-spinner class="spinner-energized"></ion-spinner>

B.<ion-spinner icon="spiral"></ion-spinner>

C.<ion-spinner class="spinner"></ion-spinner>

D.<ion-spinner class=" duration "></ion-spinner>

2．加载指示器的样式通过（　　）属性来设置加载效果显示在其他内容的顶部 。

A.z-index　　　　　B.setTimeout　　　　　C.duration　　　　　D.spinner

3．指纹验证插件具有多种属性，下列选项属于插件属性的是（　　）。

A.userAuthRequired　　　　　　　B.disablebackup

C.userName　　　　　　　　　　D.Token

4．TypeScript 中声明 AMap 对象的方法是（　　）。

A.const AMap: any;　　　　　　　B.declare const AMap: any;

C.var AMap: any;　　　　　　　　D.declare AMap: any;

5．下列不是加载属性的是（　　）。

A.showbackdrop　　　　　　　　B.enableBackdropDismiss

C.duration　　　　　　　　　　D.dismissOnPageChange

二、填空题

1．手机设备具有多种安全模式，如 ＿＿＿＿＿、＿＿＿＿＿、＿＿＿＿＿ 等。

2．指纹验证通过设置 ＿＿＿＿＿ 属性改变语言。

3．加载指示器可以通过将 ＿＿＿＿＿ 属性设置为 true 来禁用。调用 ＿＿＿＿＿ 方法在创建后关闭加载指示器。

4．装载指示器的 SVG 旋转器的名称的属性为 ＿＿＿＿＿。

5．分享时,通过 ＿＿＿＿＿ 来设置发送的内容。

三、上机题

编写代码,实现拍照后预览图片功能。效果如下图。

项目六 "Dancer 时代"我的模块实现

通过"Dancer 时代"我的模块的实现,了解视频录制、二维码扫描、语音识别及数据存储功能的实现流程,学习视频录制插件的相关知识,掌握二维码扫描插件的实际应用,具有使用 Ionic 插件实现功能的能力。在任务实现过程中:
- 了解视频录制、二维码扫描等实现流程。
- 掌握二维码扫描插件的相关知识及应用。
- 掌握 Ionic 插件的相关知识及应用。
- 具有使用 Ionic 插件实现功能的能力。

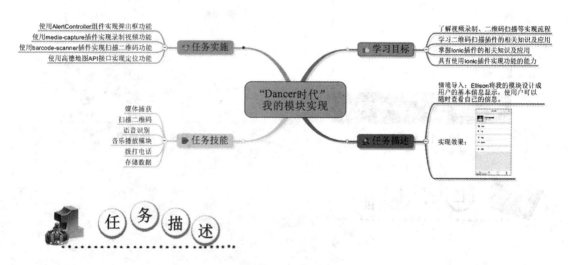

【情境导入】

Ellison 将我的模块设计成用户的基本信息显示,使用户可以随时查看自己的信息。并通过调查和了解后在该模块中添加一些手机的设置,如:视频录制、二维码扫描、手电筒、版本信息等功能。这些功能是用户在手机上经常使用的,目的是为了让用户使用软件更加便捷、舒适。本项目主要通过实现"Dancer 时代"的视频录制、二维码扫描等功能，了解 Ionic 的媒体捕获和扫描二维码。

【功能描述】

本项目将实现"Dancer 时代"我的模块界面及功能。
- 使用 AlertController 组件实现弹出框功能。
- 使用 Media Capture 插件实现录制视频功能。
- 使用 Barcode Scanner 插件实现扫描二维码功能。
- 使用高德地图 API 接口实现定位功能。

【基本框架】

基本框架如图 6.1 所示。通过本项目的学习，能将框架图 6.1 转换成效果图 6.2。

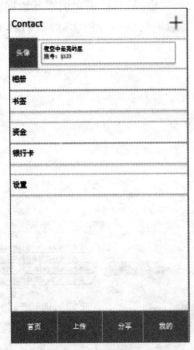 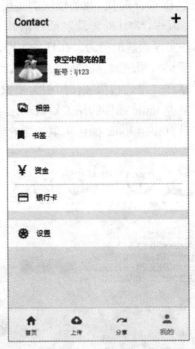

图 6.1　框架图　　　　　　　　　图 6.2　效果图

技能点 1　媒体捕获

Ionic3 提供了对音频、图片和视频采集功能的插件 - 媒体捕获插件（Media Capture），它与

Camera API 相比,不仅能获取图像,还可以录视频或者音频。实现媒体捕获首先要安装相关插件。代码如下所示。

```
ionic cordova plugin add cordova-plugin-media-capture
npm install --save @ionic-native/media-capture
```

安装后需要将插件引入项目并进行配置。在 app.module.ts 文件中引入插件,代码如下所示。

```
import { MediaCapture, MediaFile, CaptureError, CaptureImageOptions } from
'@ionic-native/media-capture';
@NgModule({
 providers: [
  StatusBar,
  SplashScreen,
  MediaCapture,
  {provide: ErrorHandler, useClass: IonicErrorHandler}
 ]
})
```

在需要使用媒体捕获插件的界面创添加击事件。代码如下所示。

```
<button id = "audioCapture" (click)="pop()"> 录制视频 </button>
```

在对应界面的 ts 文件中添加实现方法。代码如下所示。

```
pop(){
// 这里为捕获图片的方法
  let options: CaptureImageOptions = { limit: 3 };
  this.mediaCapture.captureImage(options)
    .then(
      (data: MediaFile[]) => console.log(data),
      (err: CaptureError) => console.error(err)
    );
}
```

在实现媒体捕获效果中需要用到一些属性和方法,具体的属性、方法及介绍如表 6.1 和表 6.2 所示。

表 6.1 媒体捕获属性

属 性	描 述
supportedImageModes	录制设备支持的图像大小和格式
supportedAudioModes	设备支持的录音格式
supportedVideoModes	录制视频分辨率和设备支持的格式

表 6.2 媒体捕获方法

方 法	描 述
captureAudio (options)	启动录音机应用程序并返回有关捕获的音频剪辑文件的信息。Options 参数可选有 limit（音频剪辑的最大数量，默认为 1）、Duration（音频声音片段的最大持续时间，以秒为单位）
captureImage (options)	启动相机应用程序并返回有关捕获的图像文件的信息。Options 参数可选有 limit（要捕获的最大图像数量）
captureVideo (options)	启动录像机应用程序并返回有关捕获的视频剪辑文件的信息。Options 参数可选有 limit（要记录的视频剪辑的最大数量）、Duration（每个视频剪辑的最长持续）、quality（视频质量，此参数只能与 Android 一起使用）

使用媒体捕获插件效果如图 6.3 所示。

(a)　　　　　　　　　　　(b)

图 6.3　媒体捕获插件效果图

为了实现图 6.3 的效果，新建界面，代码如 COR30601 所示。

> 代码 CORE0601　　html 文件
>
> <button id = "audioCapture" (click)="pop()"> 录制视频 </button>
> // 省略部分代码

在对应界面的 ts 文件中添加实现方法。代码如 CORE0602 所示。

> 代码 CORE0602　　ts 文件
>
> ```
> import { Component } from '@angular/core';
> import { NavController } from 'ionic-angular';
> import { MediaCapture, MediaFile, CaptureError, CaptureVideoOptions } from '@onic-native/media-capture';
> @Component({
>   selector: 'page-about',
>   templateUrl: 'about.html'
> })
> export class AboutPage {
>   constructor(public navCtrl: NavController, private mediaCapture: MediaCapture) {
>   }
>   pop(){
>     let options: CaptureVideoOptions = { limit: 3 };
>     this.mediaCapture.captureVideo(options)
>       .then(function(videoData: MediaFile[]){
>         alert(videoData[0].fullPath);
>       },
>       (err: CaptureError) => {
>         alert(err)
>       }
>     );
>   }
> }
> ```

技能点 2　扫描二维码

扫描二维码插件可用于调用系统设备的摄像头，并自动扫描条形码，获取条形码的数据并将数据返回。实现扫描二维码首先要安装相关插件，在项目目录下打开命令窗口，输入如下命令。

```
ionic cordova plugin add phonegap-plugin-barcodescanner
npm install --save @ionic-native/barcode-scanner
```

安装后需要将插件引入项目并进行配置。在 app.module.ts 文件中引入插件，代码如下所示。

```
import { BarcodeScanner } from '@ionic-native/barcode-scanner';
@NgModule({
 providers: [
   StatusBar，
   SplashScreen，
   BarcodeScanner，
   {provide: ErrorHandler, useClass: IonicErrorHandler}
 ]
})
```

在需要使用扫描二维码插件的界面添加点击事件，用于调出摄像头扫描二维码。代码如下所示。

```
<button ion-button (click)="scan()" > 点我扫描 </button>
```

在对应界面的 ts 文件中添加实现方法。代码如下所示。

```
scan (){
 this.barcodeScanner.scan().then((barcodeData) => {
 // 调出成功
 }, (err) => {
   // 调出失败
 });
}
```

在实现扫描二维码效果中需要用到 scan(options) 方法，其含义是打开条码扫描器。具体参数介绍如表 6.3 所示。

表 6.3 scan() 方法参数

参　数	类　型	描　述
options	BarcodeScannerOptions	选择传递给扫描仪的选项

使用扫描二维码插件效果如图 6.4 所示。

项目六 "Dancer 时代"我的模块实现 229

（a）

（b）

图 6.4 扫描二维码插件效果图

为了实现图 6.4 的效果，新建界面，代码如 CORE0603 所示。

代码 CORE0603　html 文件

`<button ion-button block color="secondary" class="Scan-button"`
`(click)="scan()" [disabled]="loading">` 点我扫描 `</button>`
// 省略部分代码

在对应界面的 ts 文件中添加实现方法。代码如 CORE0604 所示。

代码 CORE0604　ts 文件

`import { Component } from '@angular/core';`
`import { NavController } from 'ionic-angular';`
`import { BarcodeScanner } from '@ionic-native/barcode-scanner';`
`@Component({`
　`selector: 'page-about',`
　`templateUrl: 'about.html'`
`})`
`export class AboutPage {`
　`constructor(public navCtrl: NavController, private barcodeScanner: BarcodeScanner)`
`{`
　`}`
　`scan(){`

```
    this.barcodeScanner.scan().then((barcodeData) => {
      调出成功
      alert(barcodeData)
    }, (err) => {
      调出失败
    });
  }
}
```

技能点 3　语音识别

语音识别是将用户的语音词汇内容转化为可编辑的文字输入，通常在语音拨号、发送消息、接受语音消息等功能中使用。通过语音转换为文字进行接收与发送信息，方便用户的使用。使用 SpeechRecognition 插件即可实现语音识别，安装 SpeechRecognition 插件命令如下所示。

```
ionic cordova plugin add cordova-plugin-speechrecognition
npm install --save @ionic-native/speech-recognition
```

安装后需要将插件引入项目并进行配置。在 app.module.ts 文件中引入插件，代码如下所示。

```
import { SpeechRecognition } from '@ionic-native/speech-recognition';
//...
providers: [
  StatusBar,
  SplashScreen,
  SpeechRecognition,
  {provide: ErrorHandler, useClass: IonicErrorHandler}
]
```

在需要语音识别的界面对应的 ts 文件中添加实现方法。代码如下所示。

```
this.speechRecognition.isRecognitionAvailable()
  .then((available: boolean) => console.log(available))
// 启动识别过程
this.speechRecognition.startListening(options)
  .subscribe(
    (matches: Array<string>) => console.log(matches),
```

项目六 "Dancer 时代"我的模块实现   231

```
// 检查功能可用
  (onerror) => console.log('error:', onerror)
)
// 停止识别过程 (iOS only)
this.speechRecognition.stopListening()
// 获取语言列表
this.speechRecognition.getSupportedLanguages()
  .then(
    (languages: Array<string>) => console.log(languages),
    (error) => console.log(error)
  )
// 检查权限
this.speechRecognition.hasPermission()
  .then((hasPermission: boolean) => console.log(hasPermission))
// 请求权限
this.speechRecognition.requestPermission()
  .then(
    () => console.log('Granted'),
    () => console.log('Denied')
  )
```

提示 语音识别在微信、QQ 聊天中应用广泛，通过语音识别技能点的学习，了解到使用 Ionic 实现语音识别，扫描下方二维码，可以了解到常见的语音交互平台。快来扫我吧！

技能点 4　拨打电话

使用电话沟通具有快速、方便等特点。通过拨打电话，即使距离再远，无需见面，也能快速联系。使用 CallNumber 插件即可实现拨打电话效果。安装 CallNumber 插件命令如下所示。

```
ionic cordova plugin add call-number
npm install --save @ionic-native/call-number
```

安装后需要将插件引入项目并进行配置。在 app.module.ts 文件中引入插件，代码如下所示。

```
import { CallNumber } from '@ionic-native/call-number';
//...
providers: [
  StatusBar,
  SplashScreen,
  CallNumber,
  {provide: ErrorHandler, useClass: IonicErrorHandler}
]
```

添加点击事件,当点击时拨打号码。代码如下所示。

```
<button ion-button (click)="butt()"> 拨打电话 </button>
```

在对应的 ts 文件中添加实现方法。代码如下所示。

```
butt(){
this.callNumber.callNumber("10086", true)
    .then(() => console.log('Launched dialer!'))
    .catch(() => console.log('Error launching dialer'));}
```

使用拨打电话插件效果如图 6.5 所示。

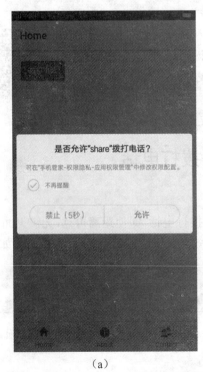

(a)　　　　　　　　　　　　(b)

图 6.5　拨打电话插件效果图

为了实现图 6.5 的效果，新建界面，代码如 CORE0605 所示。

代码 CORE0605　　html 文件

```html
<ion-header>
  <ion-navbar>
    <ion-title>Home</ion-title>
  </ion-navbar>
</ion-header>
<ion-content padding>
  <button ion-button (click)="butt()"> 拨打电话 </button>
</ion-content>
```

在对应界面的 ts 文件中添加实现方法。代码如 CORE0606 所示。

代码 CORE0606　　ts 文件

```ts
import { Component } from '@angular/core';
import { NavController } from 'ionic-angular';
import { CallNumber } from '@ionic-native/call-number';
@Component({
  selector: 'page-home',
  templateUrl: 'home.html'
})
export class HomePage {
  constructor(public navCtrl: NavController, private callNumber: CallNumber) {
  }
  butt(){
    this.callNumber.callNumber("10011", true)
      .then(() => console.log('Launched dialer!'))
      .catch(() => console.log('Error launching dialer'));
  }
}
```

技能点 5　存储数据

登录 APP 或网站时都需要输入用户名和密码，用户往往会选择"记住账号或密码"来存储这些信息，如果想要使用 Ionic 框架开发 APP 并实现该功能，需要使用数据存储来实现。数据存储的机制是数据流通过一定的格式记录在客户端上。Ionic 框架中提供了多种数据存储方式，本书主要介绍两种存储方式，分别为 localStorage 存储和 SQLite 存储。

1 localStorage 存储

在开发中,只要用户登录过一次,再次打开应用程序,都会自动填充用户名和密码,其实现是使用 localStorage 存储。localStorage 保存的数据一般是永久性保存(但当数据量过多时会自动删除),即使用户关闭当前 Web 浏览器后重新启动,数据依然存在。

在使用 localStorage 保存数据时需要用到一些方法,具体的方法及介绍如表 6.4 所示。

表 6.4 具体方法

方 法	描 述
getItem(key)	获取指定 key 本地存储的值
setItem(key, value)	将 value 存储到 key 字段
removeItem(key)	删除指定 key 本地存储的值
clear()	清除所有 localStorage 对象保存的数据

使用 localStorage 存储,在 Home 页面存储在 Contact 页面取出效果如图 6.6 所示。

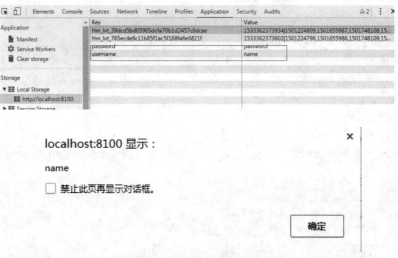

图 6.6 localStorage 存储

为了实现图 6.6 的效果,在 home.ts 文件中进行配置,代码如 CORE0607 所示。

代码 CORE0607   home.ts

```
import { Component } from '@angular/core';
import { NavController } from 'ionic-angular';
@Component({
  selector: 'page-home',
  templateUrl: 'home.html'
})
export class HomePage {
```

```
  constructor(private navCtrl: NavController) {
    // 存储 localStorage, key 值：username, value：用户名
    localStorage.setItem('username', 'name');
    // 存储 localStorage, key 值：password, value：密码
    localStorage.setItem('password', 'password');
  }
}
```

contact.ts 文件代码如 CORE0608 所示。

代码 CORE0608   contact.ts

```
import { Component } from '@angular/core';
import { NavController } from 'ionic-angular';
@Component({
  selector: 'page-contact',
  templateUrl: 'contact.html'
})
export class ContactPage {
  constructor(public navCtrl: NavController) {
//Contact 页面取出数据
    var name = localStorage.getItem('username');
    var password = localStorage.getItem('password');
    alert(name)
  }
}
```

2 SQLite 存储

通过 SQLite 也可以来存储数据，其是一款轻型的数据库。具有使用广泛、稳定、占用资源低、处理速度快等特点。和常见的客户/服务器不同，SQLite 连接到程序中，成为程序的一个主要部分，其主要通过在编程语言内用 API 调用。在使用之前需要进行安装。在项目目录下打开命令窗口，输入如下命令。

```
ionic cordova plugin add cordova-sqlite-storage
npm install --save @ionic/storage
```

安装后需要将插件引入项目并进行配置。在 app.module.ts 文件中引入插件，代码如下所示。

```
import { IonicStorageModule } from '@ionic/storage';
//...
```

```
imports: [
  BrowserModule,
  IonicModule.forRoot(MyApp),
  IonicStorageModule.forRoot()
],
```

在对应的 ts 文件中添加实现方法。代码如下所示。

```
storage.set(age 'Max');
storage.get('age').then((val) => {
  console.log('Your age is', val);
});
```

使用 SQLite 存储效果如图 6.7 所示。

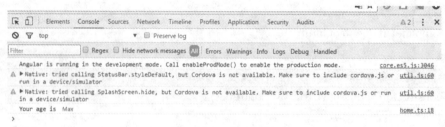

图 6.7　SQLite 存储

为了实现图 6.7 的效果，在 app.module.ts 文件中引入插件，代码如下所示。

```
import { IonicStorageModule } from '@ionic/storage';
// 省略部分代码
imports: [
  BrowserModule，
  IonicModule.forRoot(MyApp),
  IonicStorageModule.forRoot()
],
```

在对应的 ts 文件中添加实现方法。代码如下所示。

```
import { Storage } from '@ionic/storage';
// 省略部分代码
export class MyApp {
  constructor(private storage: Storage) { }
  storage.set('age', 'Max');
  storage.get('age').then((val) => {
```

项目六 "Dancer 时代"我的模块实现

```
        console.log('Your age is', val);
    });
}
```

提示 在 Ionic 开发过程中会遇到很多常见的开发问题,你是否打算放弃?扫描下面二维码,你的问题也许会迎刃而解,快来扫我吧!

通过下面九个步骤的操作,实现图 6.2 所示的"Dancer 时代"我的模块界面及功能。

第一步:切换到项目目录下,在命令窗口创建我的界面并配置(由于项目二中创建选项卡时已经创建并配置好,这里就不再进行操作)。

第二步:进行我的界面的制作。

该界面是由头部、内容区域和底部选项卡组成。通过 <ion-title> 和 <ion-icon> 标签设置头部包含的界面名称以及图标按钮。代码如 CORE0609 所示。设置样式前效果如图 6.8 所示。

```
代码 CORE0609    头部 html 代码
<ion-header>
 <ion-navbar>
  <!-- 头部 -->
  <ion-title>Contact</ion-title>
  <!-- 图标按钮 -->
  <ion-icon name="add" class="mine-icon" (click)="pop()"></ion-icon>
 </ion-navbar>
</ion-header>
```

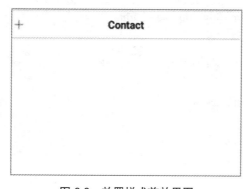

图 6.8 效置样式前效果图

设置我的界面头部样式,需要调整图标按钮的样式和位置。部分代码如CORE0610。设置样式后效果如图6.9所示。

代码CORE0610　SCSS代码

```scss
.mine-icon{
  margin-right: 3%;
  font-weight: bold;
  padding: 2%;
  position: absolute;
  right: 0;
  top: 0;
}
```

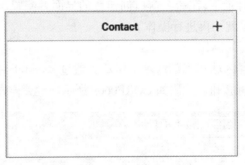

图6.9　设置样式后效果图

内容区域是由用户头像以及用户的一些信息组成。部分代码如CORE0611。设置样式前效果如图6.10所示。

代码CORE0611　html代码

```html
<ion-content class="mine-content" has-bouncing="true" overflow-scroll="false">
  <div class="mine">
  <!-- 用户头像 -->
  <img src="assets/img/logo.jpg" id="iamge">
  <!-- 用户名称 -->
  <span class="name"> 夜空中最亮的星 </span>
  <!-- 用户账号 -->
  <p>账号:<span>lj123</span></p>
  </div>
  <div class="mine-div">
```

```html
<!-- 列表 -->
  <ion-item class="mine-item">
<!-- 图标 -->
     <ion-icon name="images" class="mine-icon2"></ion-icon>
    <!-- 名称 -->
    <span> 相册 </span>
  </ion-item>
  <ion-item class="mine-item2">
    <ion-icon name="bookmark" class="mine-icon2"></ion-icon>
    <span> 书签 </span>
  </ion-item>
</div>
<div class="mine-div">
  <ion-item class="mine-item">
    <ion-icon name="logo-yen" class="mine-icon2"></ion-icon>
    <span> 资金 </span>
  </ion-item>
  <ion-item class="mine-item2">
    <ion-icon name="card" class="mine-icon2"></ion-icon>
    <span> 银行卡 </span>
  </ion-item>
</div>
<div class="mine-div" (click)="jump()">
  <ion-item class="mine-item2">
    <ion-icon name="cog" class="mine-icon2"></ion-icon>
    <span> 设置 </span>
  </ion-item>
</div>
</ion-content>
```

设置我的界面内容部分样式，需要调整图片的大小、圆角和位置，字体部分需要设置字体的样式和位置。部分代码如 CORE0612。设置样式后效果如图 6.11 所示。

图 6.10 设置样式前效果图

代码 CORE0612   SCSS 代码

```scss
//content 样式
.mine-content {
  background: #F2F2F2;
}
// 顶部 div
.mine {
  width: 100%;
  height: 80px;
  background: #ffffff;
  margin-top: 5%;
  padding-left: 4%;
}
// 用户头像
.mine img {
  width: 68px;
  height: 68px;
  margin-top: 6px;
  float: left;
}
```

```
// 名称
.name {
  display: block;
  float: left;
  margin-top: 5%;
  margin-left: 4%;
  font-size: 110%;
  font-weight: 600;
  width: 50%;
}
// 账号
.mine p {
  float: left;
  width: 50%;
  margin-left: 4%;
  margin-top: 5px;
  color: #868383;
}
// 列表 div
.mine-div {
  background: #fff;
  margin-top: 5%;
}
// 列表样式
.mine-item {
  padding: 0;
  border-bottom: 0.55px solid #c8c7cc;
  width: 92%;
  margin-left: 4%;
}
.mine-div .label-ios {
  margin: 0;
}
// 图标样式
.mine-icon2 {
  float: left;
  font-size: 150%;
  color: #0761e8;
```

```
    margin-left: 2%;
}
.mine-div span {
    display: block;
    float: left;
    line-height: 26px;
    margin-left: 5%;
    font-size: 90%;
}
.item-ios.item-block .item-inner {
    border: none;
}
.mine-item2 {
    padding: 0;
    width: 92%;
    margin-left: 4%;
}
```

图 6.11　设置样式后效果图

第三步：设置界面弹出框效果。当点击头部图标按钮时，弹出选择提示框。部分代码如 CORE0613，效果如图 6.12 所示。

代码 CORE0613　　ts 代码

```typescript
import { Component } from '@angular/core';
import { NavController, AlertController } from 'ionic-angular';
import { BarcodeScanner } from '@ionic-native/barcode-scanner';
import { MediaCapture, MediaFile, CaptureError, CaptureVideoOptions } from '@ionic-native/media-capture';
import { SetPage } from '../set/set';
import { LocationPage } from '../location/location';
@Component({
  selector: 'page-contact',
  templateUrl: 'mine.html'
})
export class MinePage {
  constructor(public navCtrl: NavController, public alertCtrl: AlertController, private barcodeScanner: BarcodeScanner, private mediaCapture: MediaCapture) {
  }
  path:string;
  pop(){
    let alerta = this.alertCtrl.create({
      title: ' 我的 ',
      buttons: [
        {
          text: ' 视频录制 ',
          handler: () => {
          }
        },
        {
          text: ' 二维码 ',
          handler: () => {
          }
        },
        {
          text: ' 我的位置 ',
          handler: () => {
        }},
      ]
    });
```

```
      alerta.present();
    }
  jump(){
    this.navCtrl.push(SetPage);
  }
}
```

图 6.12 效果图

第四步：视频录制功能：当出现弹出框时，点击录制视频，进入录制视频界面，录制完成后会返回该视频在手机中的路径（详见技能点 1）。实现视频录制功能代码如 CORE0614 所示。

代码 CORE0614　ts 代码

```
import { Component } from '@angular/core';
import { NavController, AlertController } from 'ionic-angular';
import { MediaCapture, MediaFile, CaptureError, CaptureVideoOptions } from '@ionic-native/media-capture';
@Component({
  selector: 'page-contact',
  templateUrl: 'mine.html'
})
export class MinePage {
```

```
        constructor(public navCtrl: NavController, public alertCtrl: AlertController, private
mediaCapture: MediaCapture) {
    }
    path:string;
    pop(){
    let alerta = this.alertCtrl.create({
        title: ' 我的 ',
        buttons: [
          {
            text: ' 视频录制 ',
            handler: () => {
              let options: CaptureVideoOptions = { limit: 3 };
              this.mediaCapture.captureVideo(options)
                .then(function(videoData: MediaFile[]){
                   alert(videoData[0].fullPath);
                },
                 (err: CaptureError) => {
                    alert(err)
                 }
               );
             }
           },
           {
             text: ' 二维码 ',
             handler: () => {
                 }
           },
           {
             text: ' 我的位置 ',
             handler: () => {
               console.log('Buy clicked');
             }
           }
         ]
      });
      alerta.present();
    }
}
```

第五步：二维码扫描功能，通过点击二维码进入扫描界面，扫描完成后返回内容，实现扫描二维码功能代码如 CORE0615 所示。

代码 CORE0615　ts 代码

```
{
  text: ' 二维码 ',
  handler: () => {
    this.barcodeScanner.scan().then((barcodeData) => {
          alert(barcodeData)
    }, (err) => {
          })
  }
}
```

第六步：创建地图定位页面并进行配置。

第七步：当出现弹出框时，点击我的位置，进入地图界面，在地图上显示我的位置。效果如图 6.13 所示。

图 6.13　效果图

（1）打开项目中 src 文件夹下的 index.html，引入地图接口。代码如 CORE0616 所示。

代码 CORE0616　index.html 代码

```
<script src="build/polyfills.js"></script>
<script type="text/javascript" src="http://webapi.amap.com/maps?v=1.3&key=bdfc-f87277ad9856c8694e78cfa48701"></script>
<script src="build/main.js"></script>
```

其中 key 的值为申请 API 时得到的。

(2)地图定位功能的实现,代码如 CORE0617 所示。

代码 CORE0617　定位界面(location.html)

```html
<ion-content>
  <div id="container" style="height: 500px;" ></div>
</ion-content>
```

对应 ts 代码如 CORE0618 所示。

代码 CORE0618　定位界面 ts 代码

```typescript
import { Component } from '@angular/core';
import { NavController } from 'ionic-angular';
declare const AMap: any;// 声明
@Component({
  templateUrl: 'location.html'
})
export class LocationPage {
  public map: any;
  loadMap() {
    this.map = new AMap.Map('container', {
      resizeEnable: true,
      zoom: 8,
      center: [116.39, 39.9]
    });
  }
  constructor(public navCtrl: NavController) { }
  ionViewDidLoad() {
    // 初始化地图
    let map = new AMap.Map('container', {
      view: new AMap.View2D,
      resizeEnable: true,
      zoom: 11,
      center: [116.397428, 39.90923]
    });
    AMap.plugin(['AMap.ToolBar'], function () {
      map.addControl(new AMap.ToolBar());
    })
```

```
            let marker = new AMap.Marker({
                position: map.getCenter(),
                draggable: true,
                cursor: 'move'
            });
            marker.setMap(map);
            // 设置点标记的动画效果,此处为弹跳效果
            marker.setAnimation('AMAP_ANIMATION_BOUNCE');
        }
    }
```

第八步:创建设置页面并进行配置。

第九步:进行界面跳转的实现,当点击设置时,页面发生跳转并进入到设置界面。部分代码如 CORE0619 所示。

代码 CORE0619　　ts 代码

```
import { Component } from '@angular/core';
import { NavController, AlertController } from 'ionic-angular';
import { SetPage } from '../set/set';
@Component({
  selector: 'page-contact',
  templateUrl: 'mine.html'
})
export class MinePage {
  constructor(public navCtrl: NavController, public alertCtrl: AlertController) {
  }
  jump(){
    this.navCtrl.push(SetPage);
    // 跳转到设置界面
  }
}
```

至此,"Dancer 时代"我的模块界面及功能基本完成。

本项目通过"Dancer 时代"我的模块的学习,对视频录制、二维码扫描、语音识别及数据存储功能的实现流程具有初步了解,对视频录制插件的相关知识有所认识,同时掌握了二维码扫描插件的实际应用。

speechrecognition	语音识别	LocalStorage	本地存储
capture	捕获	removeItem	组合框
mediaFile	媒体文件	scan	扫描
constructor	构造函数	dialer	拨号器
subscribe	订阅		

一、选择题

1. 下面属于媒体捕获属性是（　　）。
   A.supportsImageModes　　　　　　　　B.supportsAudioModes
   C.supportedvideoModes　　　　　　　　D.supportedVideoModes
2. 在设置扫描二维码中，需要在 app.module.ts 文件中的 providers 手动配置（　　）。
   A.StatusBar　　　　　　　　　　　　　B.SplashScreen
   C.BarcodeScanner　　　　　　　　　　D.barcodescanner
3. 下面对语音识别方法解释错误的是（　　）。
   A.isRecognitionAvailable()　检查权限　　B.startListening()　启动识别过程
   C.requestPermission()　请求权限　　　　D.getSupportedLanguages()　获取语言列表
4. 实现拨打电话功能时，需要在对应的 ts 文件 constructor 中配置（　　）。
   A.private callNumber: CallNumber　　　B.public callNumber: CallNumber
   C.public callNumber: Callnumber　　　　D.private callNumber: callNumber
5. 下面对 localStorage 存储方法解释错误的（　　）。
   A.getItem(key) 获取指定 key 本地存储的值
   B.setItem(key，value) 将 value 存储到 key 字段
   C.removeItem(key) 删除指定 key 本地存储的值
   D.clear() 清除部分 localStorage 对象保存的数据

二、填空题

1. 媒体捕获具有 _____、_____、_____ 方法。
2. 要实现扫描二维码效果，需要用到 _____、打开条码扫描器。
3. 实现语音识别停止识别过程的方法是 _____。
4. 设置拨打电话功能时，在对应的 ts 文件中添加 _____ 方法。
5. 数据存储中如何在 Contact 页面取出数据 _____。

三、上机题

在任务实施的基础上把用户的信息用 SQLite 存储起来，并通过 SQL 语句查询相关的内容。

项目七　Ionic 服务器模拟环境搭建

通过 Ionic 服务器模拟环境搭建的学习，了解 Ionic 程序和后台交互的流程，学习 Ionic 服务器中 Postman、Express、MonogoDB 的安装及使用，掌握 MonogoDB 的使用，具有搭建 Ionic 服务器环境的能力。在任务实现过程中：

- 了解 Ionic 程序和后台交互的流程。
- 学习 Postman、Express、MonogoDB 的安装及使用。
- 掌握 MonogoDB 的使用。
- 具有搭建 Ionic 服务器环境的能力。

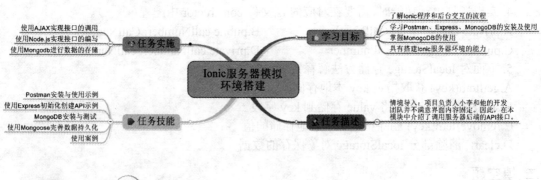

【情境导入】

至此，以舞蹈为背景的"Dancer 时代"界面设计大体完成。但项目负责人 Richard 和他的开发团队并不满意界面内容固定。因此，在本模块中介绍了调用服务器后端的 API 接口。通过 API 接口与数据库相连，及时更改数据，做到界面内容实时更新，极大程度地吸引用户。本项目主要通过实现"Dancer 时代"与后台数据交互学习 MongoDB 数据库安装与测试。

项目七 Ionic 服务器模拟环境搭建

【功能描述】

本项目将实现项目登录、注册界面与后台的交互。
- 使用 AJAX 实现接口的调用。
- 使用 Node.js 实现接口的编写。
- 使用 MongoDB 进行数据的存储。

APP 的开发避免不了与后台保持数据交互,要想实现数据交互(后端采用 Node 进行开发)需要使用 MEAN 架构,其架构是目前在互联网企业中流行的一种开发架构,旨在精简开发流程,提升开发效率,其中 MEAN 是由 MongoDB、Express、Angular、Node.js 首字母组成。该模块主要讲解使用 MongoDB、Express 这两个组件为 APP 提供后端的 API 接口,使用 Postman、Mongoose 来完善或简化 APP 应用的后端 API 服务。下面简单介绍一下这四个组件:
- Postman 是一种网页调试与发送网页 HTTP 请求的 Chrome 插件。
- Express 是一个 Web 应用框架,提供组件和模块帮助建立一个网站应用。
- MongoDB 是一个使用 JSON 形式存储数据的数据库。
- Mongoose 是一个将 JavaScript 对象与数据库产生关系的框架。

技能点 1 Postman 安装与使用示例

在项目开发中调用接口时需提前测试一下,这样就需要有一个比较给力的 Http 请求模拟工具,接口在调用阶段需要一些方便快捷的工具进行测试。目前接口请求工具五花八门,有浏览器插件型的,如 Firefox 上的 Poster 插件, Chrome 上的 Postman 插件;有工具界面型的,如 Jmeter 等。本书主要介绍 Chrome 上的 Postman 插件。

Postman 是 Google 开发的一种网页调试与发送网页 HTTP 请求的 Chrome 插件。可以方便的模拟 GET、POST 或者其他方式的请求来调试接口,并且请求头中可以附带 Headers 信息。

1 简介

一般来说,所有的 HTTP Request 都分成 URL、Headers、Method 和 Request Body 四个部分。而 Postman 针对这几部分都有针对性的工具界面。

(1)URL

在 Postman 里面输入 URL,点击 Params 按钮,会弹出一个键值编辑器,可以输入 URL 的键值对参数,Postman 会自动加入到 URL 当中。具体工具界面如图 7.1 所示。

图 7.1　URL 工具界面效果图

（2）Headers

点击 Headers 按钮，Postman 同样会弹出一个键值编辑器。可以随意添加想要的 Headers。具体工具界面如图 7.2 所示。

图 7.2　Headers 工具界面效果图

（3）Method

Method 定义请求方式，Postman 支持所有的 Method，当选择不同的请求方式，Postman 的 request body 编辑器会根据选择自动发生改变。下面主要了解一下 GET 和 POST 请求的区别：

- GET 使用 URL 传参，而 POST 将数据放在 Body 中。
- GET 的 URL 会有长度上的限制，而 POST 的数据则可以非常大。
- POST 比 GET 安全，因为 POST 请求的数据在地址栏上不可见。
- 一般 GET 请求用来获取数据，POST 请求用来发送数据。

具体工具界面如图 7.3 所示。

图 7.3　Method 工具界面

(4) Request Body

如果创建的 Request 是 POST 方式,则需要编辑 Request Body,Postman 根据 Body Type 的不同,提供了 4 种编辑方式:

● form-data 是 Web 表单默认的传输格式,编辑器允许通过设置 key-value 形式的数据来模拟填充表单。
● urlencoded 是一种编码格式,同样可以通过设置 key-value 的方式作为 URL 的参数。
● raw 可以设置常用的 JSON 和 XML 数据格式。
● binary 可以发送视频、音频、文本等文件。

具体工具界面如图 7.4 所示。

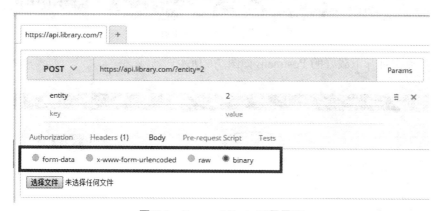

图 7.4　Request Body 工具界面

点击 Send 即可提交请求,查看请求结果时,可以以 Pretty、Raw、Preview 三种方式查看,其区别如下所示。

● Pretty 方式,可以让 JSON 和 XML 的响应内容显示的更美观规整。
● Raw 方式,显示最原始的数据。
● Preview 方式,可以把 HTML 页面自动解析显示出来。

HTTP 状态码:每发出一个 HTTP 请求之后,就会有一个响应,HTTP 本身会有一个状态码,来标示这个请求是否成功,常见状态码如下。

● 200,2 开头的都表示这个请求发送成功,最常见的就是 200。
● 300,3 开头的代表重定向,最常见的是 302 。
● 400,400 代表客户端发送的请求有语法错误,401 代表访问的页面没有授权,403 代表没有权限访问这个页面,404 代表没有这个页面。
● 500,5 开头的代表服务器有异常,500 代表服务器内部异常,504 代表服务器端超时,没返回结果。

2　安装

第一步:打开官网(https://www.getpostman.com)下载 Postman 插件,如图 7.5 所示。

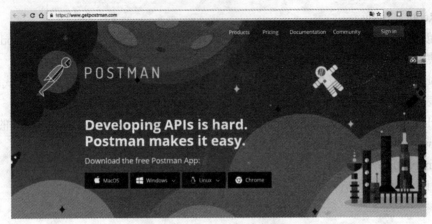

图 7.5 Postman 插件主界面

第二步:点击打开下载好的 Postman,如图 7.6 所示。

图 7.6 Postman 界面

第三步:在里面输入需要测试的接口地址,选择请求方式,在下面手动添加相应的键值,点击 send,可以看到相关的响应信息,如图 7.7 所示。

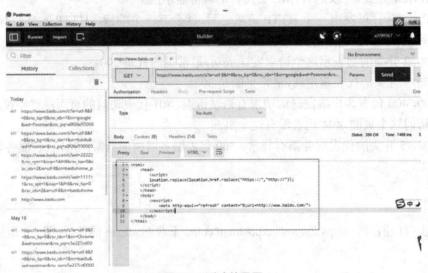

图 7.7 测试效果图

3 使用说明

（1）发送一个 GET 请求

比如向百度发送一个搜索请求。进入 Postman 界面中，选择 GET 请求，在输入框中输入百度网址。输入完成之后，点击 send。如果测试成功，如图 7.8 所示。从图中看到，请求的状态码 Status 是 200，表示此次请求发送成功。本次的请求响应时间是 243 ms，而在下列的是返回的百度首页的元素信息。这样百度首页接口测试就成功了。具体界面如图 7.8 所示。

图 7.8　GET 请求测试

（2）发送 POST 请求

POST 表单提交，需要设置请求方法、URL、参数等属性。当选择 x-www-form-urlencoded 的参数方式后，Postman 自动的设置了 Content-Type，不需要人工干预，这就是使用一款流行工具的好处。具体界面如图 7.9 所示。

图 7.9　POST 请求测试

提示 无论生活中遇到多大困难,只要你拥有梦想,你就能克服所有困难。扫描图中二维码,我们来看看没有翅膀也能飞翔的杰西卡·考克斯故事。快来扫我吧!

技能点 2　使用 Express 初始化创建 API 示例

Express 是基于 Node.js 平台,快速、开放、极简的 Web 开发框架,提供一系列强大特性帮助创建各种 Web 应用,包含丰富的 HTTP 工具以及任意排列组合的 Connect 中间件,使创建强健、友好的 API 变得快速又简单。因此使用 Express 可以快速地搭建一个完整功能的网站。下面从安装、工程结构、工作原理简单介绍一下 Express。

1　安装

第一步:安装 Express。打开命令窗口,输入下面命令,安装效果如图 7.10 所示。

```
npm install express-generator -g
```

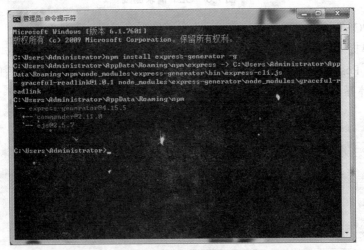

图 7.10　安装 Express

第二步:使用以下命令新建项目,项目名称为 bloo,新建项目效果如图 7.11 所示。

```
express bloo
```

第三步:进入项目并安装相关依赖,如图 7.12 所示。

```
cd bloo && npm install
```

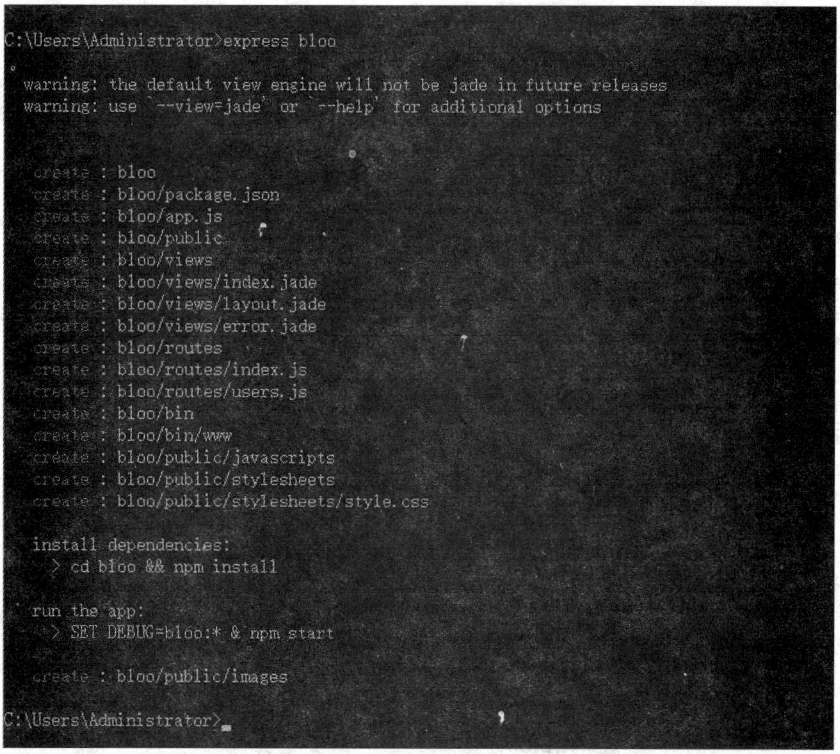

图 7.11 新建项目

图 7.12 安装依赖

第四步:启动 Express。

```
npm start
```

第五步:在浏览器中输入 http://localhost:3000/,并查看输出结果,输出结果如图 7.13 所示。

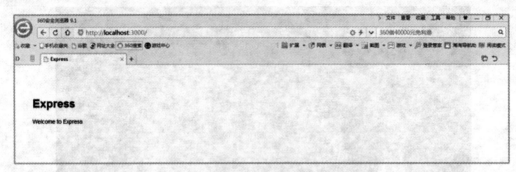

图 7.13　输出结果

2　项目结构

打开上面生成的项目，生成的项目如图 7.14 所示。

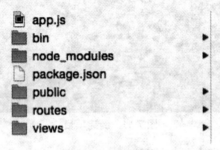

图 7.14　项目结构

- app.js：启动文件。
- package.json：存储着工程的信息及模块依赖，当在 dependencies 中添加依赖的模块时，运行 npm install，npm 会检查当前目录下的 package.json，并自动安装所有指定的模块。
- node_modules：存放 package.json 中安装的模块，当在 package.json 添加依赖的模块并安装后，存放在这个文件夹下。
- public：存放 image、CSS、JS 等文件。
- routes：存放路由文件。
- views：存放视图、模版文件。
- bin：存放可执行文件。

3　工作原理

Express 的工作原理是 routes 中存放路由文件，views 中存放视图文件，这就相当于 MVC 模式中的 C 和 V，而 index.ejs 文件中的 <%= title%> 是 ejs 模板引擎的语句，可以将后台传递来的 title 数据在页面中显示。使用 Express 大大减少了代码函数，而且逻辑更为简洁，所以使用 Express 可以提高开发效率并降低工程维护成本。Express 有几个比较重要的概念：路由、中间件和模版引擎。

（1）路由

路由（Routing）是由一个 URI 和一个特定的 HTTP 方法（GET、POST 等）组成的。每一个路由都可以有一个或者多个处理器函数,当匹配到路由时,这些函数将被执行。开发人员可以为 Web 页面注册路由,将不同的路径请求区分到不同的模块中去。路由的定义结构如下。

```
app.METHOD(path, HANDLER)
```

- app 是 Express 对象的一个实例。
- METHOD 是一个 HTTP 请求方法。
- path 是服务器上的路径。
- HANDLER 是当路由匹配时要执行的函数。函数中又有两个参数 req 和 res,代表请求信息和响应信息。

当访问主页时,调用 ejs 模板来渲染 views/index.ejs 模板文件,其中 Get 指 HTTP 的 Get 请求方式。基本的路由示例如下所示。

```javascript
var express = require('express');
var app = express();
// 对网站首页的访问返回 "Hello World!" 字样
app.get('/', function(req, res) {
  res.send('hello world');
});
```

(2) 中间件

中间件(Middleware)是一个函数,它可以访问请求对象(request object (req))、响应对象(response object (res)),处于请求 - 响应循环流程中的中间件,一般被命名为 next 的变量。

应用级中间件绑定到 app 对象使用 app.use() 和 app.METHOD(),其中 METHOD 是需要处理的 HTTP 请求的方法,例如 GET、PUT、POST 等。使用中间件代码如下所示。

```javascript
var app = express();
// 没有挂载路径的中间件,应用的每个请求都会执行该中间件
app.use(function (req, res, next) {
  console.log('Time:', Date.now());
  next();
});
// 挂载至 /user/:id 的中间件,任何指向 /user/:id 的请求都会执行它
app.use('/user/:id', function (req, res, next) {
  console.log('Request Type:', req.method);
  next();
});
// 路由和句柄函数 ( 中间件系统 ),处理指向 /user/:id 的 GET 请求
app.get('/user/:id', function (req, res, next) {
  res.send('USER');
});
```

(3)模板引擎

模板引擎是一个将页面模板和数据结合起来生成 HTML 页面的工具。通过模板引擎,可以在 HTML 文件中直接使用后台传递过来的数据,大大提高了开发效率。

使用之前先安装相应的模板引擎 npm 软件包。-e 是指定 ejs 作为模板引擎,而默认的模板引擎是 jade。创建一个 ejs 作为模板引擎的项目代码如下。

```
express -e blog(项目名称)
```

需要在应用中进行如下设置才能让 Express 渲染模板文件,设置模板文件的存储位置和使用的模板引擎。代码如下所示。

```
app.set('views', path.join(__dirname, 'views'));
app.set('view engine', 'ejs');
```

编写 views 目录下名为 index.ejs 的 ejs 模板文件,内容如下所示。

```
<!DOCTYPE html>
<html>
<head>
  <title><%= title %></title>
  <link rel='stylesheet' href='/stylesheets/style.css' />
</head>
<body>
<h1><%= title %></h1>
<p> 姓名:<%= title %>  </p>
</body>
</html>
```

然后创建一个路由渲染 index.ejs 文件。在 routers/index.js 中通过 res.render() 渲染模板。它接收两个参数,第一个是模板名称,即 views 目录下的模板文件名;第二个参数是传递给模板的数据对象。

举例:当代码为 res.render('index', {title: 'Express'}) 时,模板引擎会把 <%= title %> 替换为 Express,然后把替换后的页面展示给用户。此时向主页发送请求,会被渲染为 HTML 具体代码如下所示。

```
router.get('/', function(req, res, next) {
  myModel.findOne({name:"zhangangs"}, function (err, user) {
    console.log(user);
    res.render('index', {title: 'Express', user: user });
  });
});
```

技能点 3　MongoDB 安装与测试

1　简介

通过使用数据库可以保存网站的用户、业务等数据。本书采用 MongoDB 数据库（基于文档的非关系型数据库（NoSQL）），其使用集合（相当于表）和文档（相当于行）来描述和存储数据。该数据库支持的查询语言非常强大，几乎实现类似关系数据库单表查询的绝大部分功能，且支持对数据建立索引，其由 C++ 语言编写。MongoDB 数据库旨在为 Web 应用提供可扩展的高性能数据存储解决方案，它在多种场景下可用于替代传统关系型数据库的键 / 值存储方式。具有高性能、易部署、易使用、易存储等特点。

2　MongoDB 的安装和服务的启动

第一步：下载 msi 文件。MongoDB 数据库提供了可用于 32 位和 64 位系统的预编译二进制包，通过从 MongoDB 数据库官网（https://www.mongodb.com/download-center#community）下载安装。下载 MongoDB 及可选的版本如图 7.15 所示。

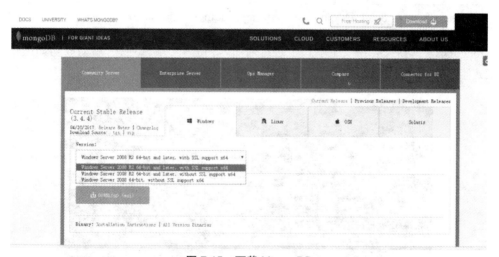

图 7.15　下载 MongoDB

● MongoDB for Windows 64-bit：适合 64 位的 Windows Server 2008 R2、Windows 7 及最新版本的 Window 系统。

● MongoDB for Windows 32-bit：适合 32 位的 Window 系统及最新的 Windows Vista。

● MongoDB for Windows 64-bit Legacy：适合 64 位的 Windows Vista、Windows Server 2003 及 Windows Server 2008。

第二步：下载后双击该文件，按提示安装即可。安装过程中，通过点击 "Custom(自定义)" 按钮更改安装目录，默认存放在 C:\Program Files\MongoDB，安装 MongoDB 效果如图 7.16 所示。

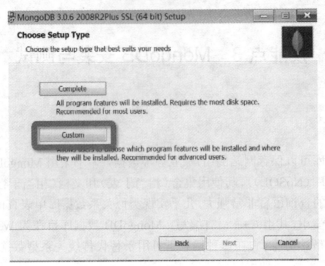

图 7.16 安装 MongoDB

第三步：指一个文件夹存放数据，在之前新建的 bloo 目录下新建 data 目录，它是存放数据的文件夹。

第四步：配置 MongoDB 服务端。打开 cmd 命令行，进入安装 MongoDB 的 bin 目录下，输入如下命令启动 MongoDB 服务。

```
mongod.exe --dbpath D:\blog\data --storageEngine=mmapv1
```

MongoDB 启动后，会在一个端口上进行监听，等待客户端来连接，启动成功如图 7.17 所示，默认监听的端口是 27017。

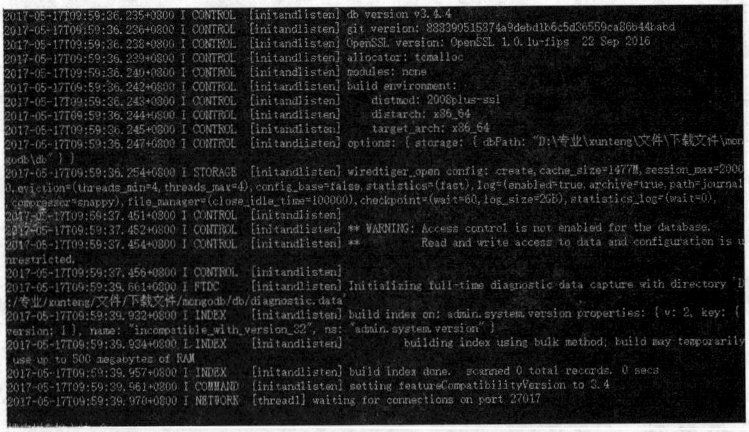

图 7.17 启动 MongODB 服务

并且 data 文件夹里会新出现一些文件及文件夹，用来存储数据库中的数据，data 文件夹结构如图 7.18 所示。

图 7.18 data 文件夹结构

第五步:在命令窗口输入以下命令。mongo Shell 默认连接到了 test 数据库。使用 db 命令查看当前操作的数据库,如图 7.19 所示。

```
mongo
db
```

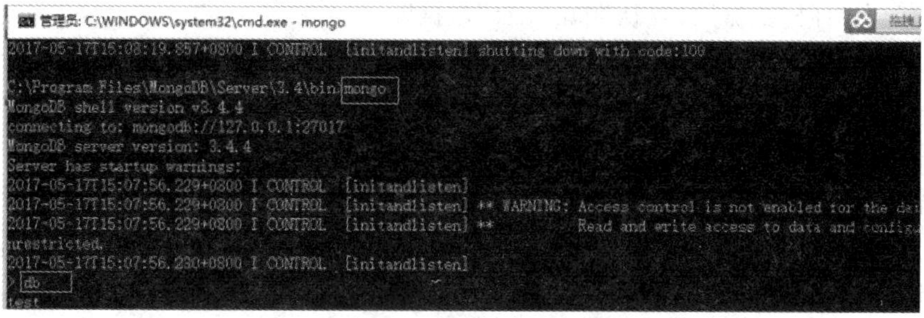

图 7.19 连接及查看数据库

第六步:在浏览器输入 http://localhost:27017/,MongoDB 启动成功效果如图 7.20 所示。

图 7.20 MongoDB 启动成功

3 数据库管理

(1)数据库操作

● 创建数据库。通过 use+"数据库名"创建数据库,使用"db"查看新建的库。新建数据库 use blog(此时输入 db 不会显示 blog,需要进行插入数据等操作,才会被记录到本地)。操作数据库效果如图 7.21 所示。

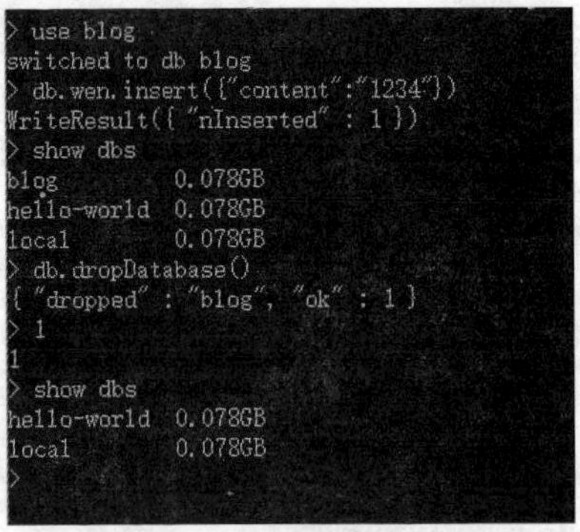

图 7.21 操作数据库效果图

- 添加数据。在 blog 中插入数据：db.post.insert({"content": "123"})。此时命令行中出现 WriteResult({"nInserted" : 1})，插入数据成功，同时在数据存放目录（blog\data）中出现新增的 blog.ns 和 blog.0 文件。
- 删除数据库。db.dropDatabase() 命令用于删除现有的数据库。当删除选定的数据库时，如果没有选择任何数据库，默认删除"test"数据库。
- 查看数据库。通过使用 show dbs（默认含有名为 test 的数据库）查看本地所含数据库
- 查看数据库名。通过使用 db 查看当前操作的数据库名。

（2）集合操作

数据库具有不同的集合，但操作方式大体相似，下面针对某个集合进行相关操作。

- 创建集合：通过 db.createCollection('col') 创建一个名为 col 的集合，或通过 db.col.insert({}) 创建（向 col 集合里插入空文档，当无 col 集合时，会自动创建一个 col 集合）。创建 col 集合，代码如下所示。

>db.col.insert({title: 'MongoDB 教程 ',
　　description: 'MongoDB 是一个 NoSQL 数据库 ',
　　url: 'http://www.runoob.com',
　　tags: ['mongodb', 'database', 'NoSQL'],
　　likes: 100
})

- 查看数据库里的集合。通过 show collections 进行查看数据库里的集合。当没有集合时，不显示。当具有集合时，则显示相应的集合名词及 system.indexes。
- 删除 items 集合。通过 db.items.drop() 删除集合。

集合操作效果如图 7.22 所示。

（3）文档操作

集合由文档组成，文档具有增、删、改、查等操作，如下所示。

- 增：在 col 集合中插入一个或多个文档。插入单个文档：db.col.insert({"content":"123"}) 插入多个文档：db.col.insert([{"content":"456"},{"content":"789"}])。

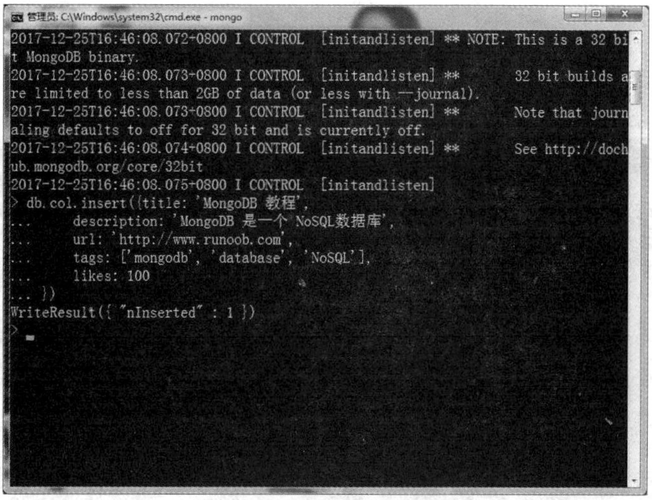

图 7.22 集合操作

- 删：删除 col 集合中的文档：db.post.remove({"content","123"})。
- 改：修改 col 集合中的文档：db.col.update({"content","123"},{$set{"content","abc"}})。
- 查：查询 col 集合中的文档。查询所有文档：db.col.find()；查询 content 为 abc 的文档：db.post.find({},{"content","abc"})。

文档操作效果如图 7.23 所示。

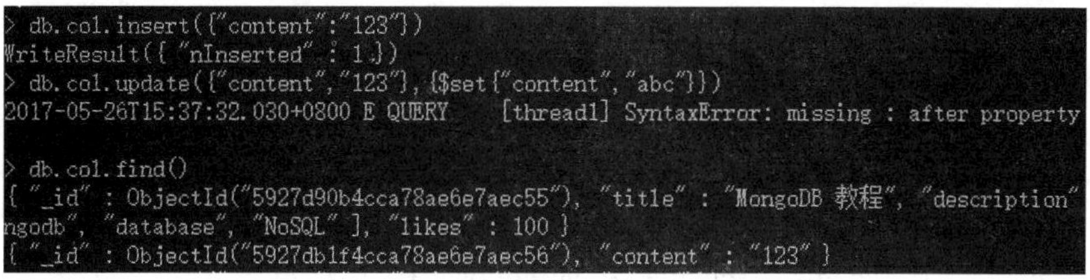

图 7.23 操作文档

技能点 4 使用 Mongoose 完善数据持久化

Mongoose 是一款易于使用的 Web 服务器，它可以嵌入其他应用程序中，为其提供 Web 接口，是一个将 JavaScript 对象与数据库产生联系的框架。使用 Mongoose 可以新建数据库、集合并对集合内的文档进行 CRUD 操作，其提供了 Schema、Model 和 Document 对象，使操作更加方便。通过 Schema 对象可定义文档结构（类似表结构），如定义字段、类型、唯一性、索引和验证等。Model 对象表示集合中的所有文档。Document 对象表示集合中的单个文档。使用 Mongoose 步骤如下所示。

第一步：安装 Mongoose。通过以下命令安装 Mongoose、Mongoose 依赖的 MongoDB driver 以及 regexp 等模块，安装效果如图 7.24 所示。

```
npm install mongoose –save
```

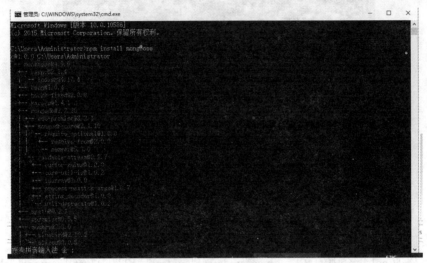

图 7.24 安装 Mongoose

第二步：修改 app.js，添加下面的代码连接数据库。

```
// 引入 Mongoose 模块
var mongoose = require('mongoose');
// 连接数据库
mongoose.connect('mongodb://localhost:27017/datas');
mongoose.connection.on('error', console.error.bind(console, ' 连接数据库失败 '));
```

第三步：定义一个 Schema。Schema 是 Mongoose 中的模型对象，与关系型数据库中的表结构类似。

```
var PersonSchema = new mongoose.Schema({
    name:String   // 定义一个属性 name，类型为 String
});
```

提示　当对 MongoDB 数据库了解后，你是否觉得文档数据库只此一种。扫描图中二维码，查看其他版本数据库。

项目的整体架构及基本界面在前面项目中已经完成，这里主要配置登录、注册页面与后台

进行的一个用户名和密码的存储、读取效果。通过下面十一个步骤的操作，实现"Dancer 时代"前端和后台的数据交互。

第一步：创建注册页面并进行相关配置。

第二步：注册页面的制作。

该界面分为三部分：输入框区域、提交按钮以及关于注册的法律知识和用户协议的链接说明。顶部是由两个输入框组成，中间是一个注册的按钮，底部是一段文字说明。部分代码如 CORE0701。设置样式前效果如图 7.25 所示。

代码 CORE0701　注册界面

```html
<ion-header>
  <ion-navbar>
    <ion-title> 注册 </ion-title>
  </ion-navbar>
</ion-header>
<ion-content  has-bouncing="true" overflow-scroll="true">
  <div class="login-mian">
    <div class="register-content">
      <ion-list>
        <ion-item>
          <ion-label floating>Username</ion-label>
          <ion-input [(ngModel)]="username" type="text" (blur)="regi()"></ion-input>
        </ion-item>
        <ion-item>
          <ion-label floating>Password</ion-label>
          <ion-input [(ngModel)]="password" type="password"></ion-input>
        </ion-item>
      </ion-list>
      <button class="rebutton" ion-button block (click)="register()">
        注册
      </button>
        <p class="text"> 点击"注册新账号"即表示您同意并遵守 lion
          <a class="link-text" href="#/userText"> 用户协议 </a> 和
          <a class="link-text" href="#/protocolText"> 隐私政策 </a>
        </p>
    </div>
  </div>
</ion-content>
```

图 7.25 注册界面

第三步:打开命令窗口切换到想要创建 node 项目的路径下,输入 express app(项目名称)创建项目,如图 7.26 所示。

图 7.26 创建 node 项目

第四步：在上述命令行中输入 cd app 切换到项目路径，之后输入 npm install 安装项目依赖。

第五步：重新打开一个命令窗口，切换到数据库安装路径的 bin 文件夹下，运行 mongod.exe --dbpath D:\blog\data（存放数据的路径）--storageEngine=mmapv1，之后再打开一个命令窗口，再次切换到上述路径下，输入 mongo 启动 MongoDB 数据库。

第六步：打开项目路径下的 routes/index.js 进行后台注册接口验证账号是否存在代码如 CORE0702 所示。

代码 CORE0702　　验证账号 ts

```
router.get('/checkuser', function(req, res, next) {
  var MongoClient = require('mongodb').MongoClient;
  var DB_CONN_STR = 'mongodb://localhost:27017/runoob';
  var selectData = function(db, callback) {
    // 连接到表 site
    var collection = db.collection('site');
    // 查询数据是否存在
    var sss=req.query.user;
    var whereStr = {"name":sss};
    collection.find(whereStr).toArray(function(err, result) {
      if(err)
      {
        console.log('Error:'+ err);
        return;
      }
      callback(result);
    });
  }
  MongoClient.connect(DB_CONN_STR, function(err, db) {
    console.log(" 连接成功！ ");
    selectData(db, function(result) {
      console.log(result);
      var ss=result.toString();
      if(ss.length==0){
        // 存在
        res.jsonp({data:"true"})
      }else {
```

```
           res.jsonp({data:"false"})
       }
       db.close();
     });
   });
 });
```

第七步：注册接口存储账号密码，代码如 CORE0703 所示。

代码 CORE0703　　注册接口存储账号密码 ts 文件

```
router.get('/register', function(req, res, next) {
 var MongoClient = require('mongodb').MongoClient;
 var DB_CONN_STR = 'mongodb://localhost:27017/runoob';
 var selectData = function(db, callback) {
  // 连接到表
  var collection = db.collection('site');
  // 查询数据是否存在
  //var ss=db.getCollection('site').find({});
  var sss=req.query.user;
  var whereStr = {"name":sss};
  collection.find(whereStr).toArray(function(err, result) {
   if(err)
   {
    console.log('Error:'+ err);
    return;
   }
   callback(result);
  });
 }
 MongoClient.connect(DB_CONN_STR, function(err, db) {
  console.log(" 连接成功！");
  selectData(db, function(result) {
   console.log(result);
   var ss=result.toString();
   if(ss.length==0){
    // 存在
    var MongoClient = require('mongodb').MongoClient;
```

```javascript
            var DB_CONN_STR = 'mongodb://localhost:27017/runoob';
         var insertData = function(db, callback) {
            // 连接到表 site
            var collection = db.collection('site');
          // 插入数据
          var user=req.query.user;
          var password=req.query.password;
          var date=new Date();
          var time=date.getTime();
          var data = [{"name":user,"password":password,"id":time}];
          collection.insert(data, function(err, result) {
            if(err)
            {
              console.log('Error:'+ err);
              res.jsonp({data:"false"});
              return;
            }
            callback(result);
          });
        }
        MongoClient.connect(DB_CONN_STR, function(err, db) {
          console.log(" 连接成功！ ");
          insertData(db, function(result) {
            console.log(result);
            res.jsonp({data:"true"})
            db.close();
          });
        });
      }else {
       res.jsonp({data:"false"})
       }
       db.close();
      });
     });
    });
```

第八步：打开命令窗口进入 Ionic 项目路径，输入 npm install jquery slick-carousel 命令引用 jQuery 插件。

第九步：用户注册时分两个部分，首先需要验证账号是否存在，之后进行账号、密码的存储，代码如 CORE0704 所示。效果如图 7.27 所示。

代码 CORE0704　数据存储 ts

```ts
import { Component } from '@angular/core';
import { NavController, NavParams } from 'ionic-angular';
import * as $ from "jquery";
import "slick-carousel";
import { LoginPage } from '../login/login';
@Component({
  selector: 'page-item-details',
  templateUrl: 'register.html'
})
export class RegisterPage {
  selectedItem: any;
  username:string;
  password:string;
  ists:string="false";
  constructor(public navCtrl: NavController, public navParams: NavParams) {
    this.selectedItem = navParams.get('item');
  }
  regi(){
    var user=this.username;
    var that=this;
    $.ajax({
      type : "get",  // 提交方式
      url : "http://localhost:3000/checkuser", // 路径
      data:{user:user}, // 数据，这里使用的是 Json 格式进行传输
      dataType:'json',
      success : function(istrue) { // 返回数据根据结果进行相应的处理
        console.log(istrue)
        if(istrue.data=="true"){
          alert(" 可以注册 ");
          that.ists="true";
        }else {
          alert(" 不可以注册 ");
        }
```

```
    },error:function(istrue){
     alert("eee");
     console.log(istrue);
    }
  });
 }
 register(){
  var user=this.username;
  var pw=this.password;
  var that=this;
  if(this.ists=="true"){
    $.ajax({
     type : "get", // 提交方式
     url : "http://localhost:3000/register", // 路径
     data:{user:user,password:pw}, // 数据，这里使用的是 Json 格式进行传输
     dataType:'json',
     success : function(istrue) { // 返回数据根据结果进行相应的处理
       console.log(istrue)
       if(istrue.data=="true"){
        alert(" 注册成功 ");
        that.navCtrl.push(LoginPage);
       }else {
        alert(" 注册失败 ");
       }
    },error:function(istrue){
     alert("eee");
     console.log(istrue);
    }
   });
  }
 }
}
```

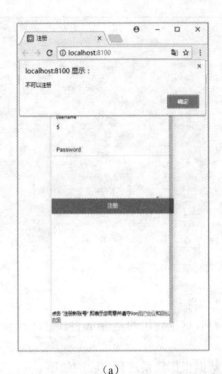
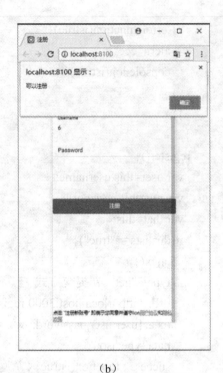

（a）　　　　　　　　　　　　　　（b）

图 7.27　验证数据是否存在

第十步：编写登录接口。代码如 CORE0705 所示。

代码 CORE0705　　登录接口

```
router.get('/login', function(req, res, next) {
  var MongoClient = require('mongodb').MongoClient;
  var DB_CONN_STR = 'mongodb://localhost:27017/runoob';
  var selectData = function(db, callback) {
    // 连接到表
    var collection = db.collection('site');
    // 查询数据
    //var ss=db.getCollection('site').find({});
    var user=req.query.user;
    var password=req.query.password;
    var whereStr = {"name":user,"password":password};
    collection.find(whereStr).toArray(function(err, result) {
      if(err)
      {
        console.log('Error:'+ err);
        return;
      }
```

```
    callback(result);
  });
}
MongoClient.connect(DB_CONN_STR, function(err, db) {
  console.log(" 连接成功！");
  selectData(db, function(result) {
    console.log(result);
    var ss=result.toString();
    if(ss.length==0){
      res.jsonp({data:"false"})
    }else {
      res.jsonp({data:"true"})
    }
    db.close();
  });
});
});
```

第十一步：创建登录页面（项目一已经创建过了，这里就不在进行创建），编写登录时与后台进行交互的代码，代码如 CORE0706 所示。效果如图 7.28 所示。

代码 CORE0706　界面交互

```
import { Component } from '@angular/core';
import { NavController } from 'ionic-angular';
import { TabsPage } from '../tabs/tabs';
import * as $ from "jquery";
import "slick-carousel";

@Component({
  selector: 'page-list',
  templateUrl: 'login.html'
})
export class LoginPage {
  username:string;
  // 定义 username 类型
  password:string;
  constructor(public navCtrl:NavController) {}
  login() {
```

```
    var user = this.username;
    // 给 user 变量赋值
//this.username 为 input 中的 value 值
    var pw = this.password;
    // 给 pw 变量赋值
////this.password 为 input 中的 value 值
    var that=this;
    $.ajax({
     type : "get", // 提交方式
     url : "http://localhost:3000/login", // 路径
     data:{user:user,password:pw}, // 数据,这里使用的是 Json 格式进行传输
     dataType:'json',
     success : function(istrue) { // 返回数据根据结果进行相应的处理
      // alert(istrue.msg);
      console.log(istrue)
      if(istrue.data=="true"){
        alert(" 登录成功 ");
        that.navCtrl.push(TabsPage);
       }else {
        alert(" 登录失败 ");
       }
     },error:function(istrue){
      alert("eee");
      console.log(istrue);
     }
    });
   }
  }
```

(a)   (b)

图 7.28   登录验证

至此,"Dancer 时代"前端和后台的数据交互基本完成。

本项目通过"Dancer 时代"前端和后台的数据交互的实现,对 Ionic 程序和后台交互的流程具有初步的了解,并详细了解 Ionic 服务器中 Postman、Express、MonogoDB 的安装及使用,具有了搭建 Ionic 服务器环境的能力。

server	服务员	attribute	属性
binary	二进制	postman	邮递员
method	方式	pretty	漂亮的
jmeter	测试	request	请求
preview	未加工		

一、选择题

1. 使用之前先安装相应的模板引擎 npm 软件包。-e 是指定（　　）作为我们的模板引擎。
   A.HTML　　　　B.Express　　　　C.jade　　　　D.ejs

2. MongoDB 是基于文档的非关系型数据库（NoSQL），其使用（　　）来描述和存储数据。
   A. 集合和文档　　B. 键/值存储方式　　C. 集合和表　　D. 集合和行

3. MongoDB 是由（　　）语言编写,旨在为 WEB 应用提供可扩展的高性能数据存储解决方案。
   A.Java　　　　B.C++　　　　C.php　　　　D.Html

4. 使用 mongoose 可以新建数据库、新建集合、对集合内的文档进行 CRUD 操作,提供了（　　）、Model 和 Document 对象,用起来更为方便。
   A.Query　　　　B.Aggregate　　　　C.driver　　　　D.Schema

5. Express 创建的工程,存储着工程的信息及模块依赖（　　）。
   A.package.json　　B.routes　　　　C.public　　　　D.views

二、填空题

1. MEAN 是 _____、Express、Angular、_____ 首字母缩写,全都是用 Javascript 描述的,因此有人称它是 Javascript 开发工程师全栈的开发框架。

2. MongoDB 是一个使用 _____ 形式存储数据的数据库。

3. MongoDB 在许多场景下可用于替代传统的关系型数据库或 _____。

4. 目前接口请求工具五花八门,有浏览器插件型的,如 firefox 上的 Poster 插件,chrome 上的 _____ 插件,有工具界面型的,如 Jmeter 等。

5. Express 生成工程的启动文件为 _____。

三、上机题

使用 Nodejs 编写符合以下要求的代码。要求：
1. 使用 Express 创建 node 项目；
2. 链接 MongoDB 数据库；
3. 使用 Nodejs 写一个增加数据接口。

项目八 "Dancer 时代"发布

通过发布"Dancer 时代"项目,了解 Ionic 项目在 Android 及 IOS 手机下的发布流程,学习 Ionic 环境如何签名和优化 APK,掌握如何发布 Ionic 项目,具有能够根据需求分析和设计制作并发布项目的能力,在任务实现过程中:

- 了解 Ionic 项目在 Android 及 IOS 手机下的发布流程。
- 掌握如何发布 Ionic 项目。
- 具有能够根据需求分析和设计制作并发布项目的能力。

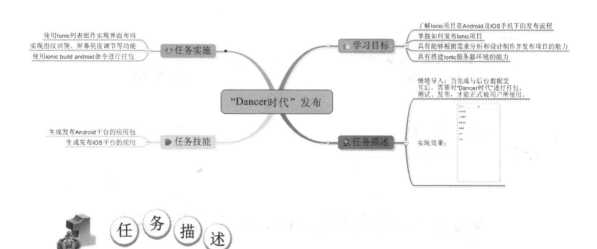

【情境导入】

当完成与后台数据交互后,需要对"Dancer 时代"进行打包、测试、发布,才能正式被用户所使用。目前手机版本主要为 Android 和 IOS,因此,项目负责人 Richard 决定将"Dancer 时代"生成 Android 和 IOS 应用包并进行发布。本项目主要通过对"Dancer 时代"打包进行了解和操作,做出一个属于自己的 APP 应用。

【功能描述】

本项目将发布和部署"Dancer 时代"。
- 使用 Ionic 列表组件实现界面布局。
- 使用 Ionic 插件实现指纹识别、屏幕亮度调节等功能。
- 使用 ionic build android 命令进行打包。

【基本框架】

基本框架如图 8.1 所示。通过本项目的学习,能将框架图 8.1 转换成效果图 8.2。

图 8.1 框架图　　　　　　　图 8.2 效果图

技能点 1　生成发布 Android 平台的应用包

1　打包 APK

Ionic 功能开发完成后,就需要将项目打包发布到应用市场。有两种打包的版本:一种是 debug 调试版,另一种是 release 发布版。打包 debug 调试版需在工程目录下执行 ionic build android;打包 release 发布版,在工程目录下执行以下命令。

```
ionic build android --release
```

以上命令会在 platforms\android\build\outputs\APK 目录下生成未签名的 release 应用包（android-release-unsignedAPK）。打包 debug 调试版，里面包含签名，可以直接在手机上安装，但不能发布到应用市场，想要发布到应用商店，需要重新进行签名。

2 签名 APK

APK 签名就是给 APP 一个唯一身份，Android APP 那么多，拥有 APK 签名才可以上传到 APP 商店中，升级应用时若包名一致但签名不一致，APP 将会安装失败，所以正式版 APP 都需要签名。

（1）生成签名的秘钥

生成签名的秘钥需要使用 JDK 附带的 keytool 工具，该文件一般在 JDK 路径下的 bin 目录下。使用如下命令生成签名文件，效果如图 8.3 所示。

```
keytool -genkey -v -keystore android.keystore -alias android.keystore -keyalg RSA -keysize 2048 -validity 10000
```

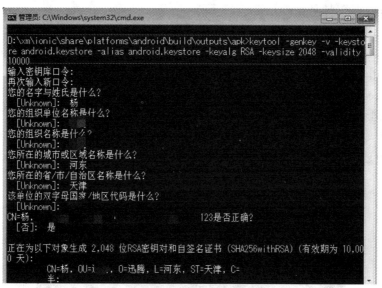

图 8.3 生成签名的秘钥

其中 keytool 是工具名称，-genkey 意味着执行的是生成数字证书操作，-v 表示将生成证书的详细信息打印出来。

- -keystore android.keystore 证书的文件名。
- -alias android.keystore 表示证书的别名。
- -keyalg RSA 生成密钥文件所采用的算法。
- -validity 10000 该数字证书的有效期。

注意：请记住输入的密码，进行签名时需要验证，生成完成后，就可以看到目录下有一个 android.keystore 文件,该文件必须妥善保存,如果丢失,将无法向应用程序提交更新。如图 8.4 所示。

图 8.4　验证完成效果图

(2)进行签名

对 APK 文件签名需要使用 JDK 附带的 jarsigner 工具,该文件一般在 JDK 路径下的 bin 目录下,进入生成 APK 文件的目录下,在命令行内输入如下命令。

> jarsigner -verbose -keystore Android.keystore -signedjar enhanced_signed.apk enhanced.apk android.keystore

其中:
- jarsigner 是工具名称。
- -verbose 表示将签名过程中的详细信息打印出来。
- -keystore Android.keystore 表示签名证书文件,这里需要替换成自己的 keystore。
- -signedjar enhanced_signed.apk 表示签名后生成的 APK 名称。
- enhanced.apk 需要签名的 APK 名称。
- android.keystore 表示 keystore 的别名,这里需要替换成自己的 keystore 的别名。

执行过程中会要求你输入密码,请输入生成密钥文件时录入的密码。执行完成后,该 APK 应用包就已经签名了,可以将这个 APK 对外发布。签名过程如图 8.5 所示。

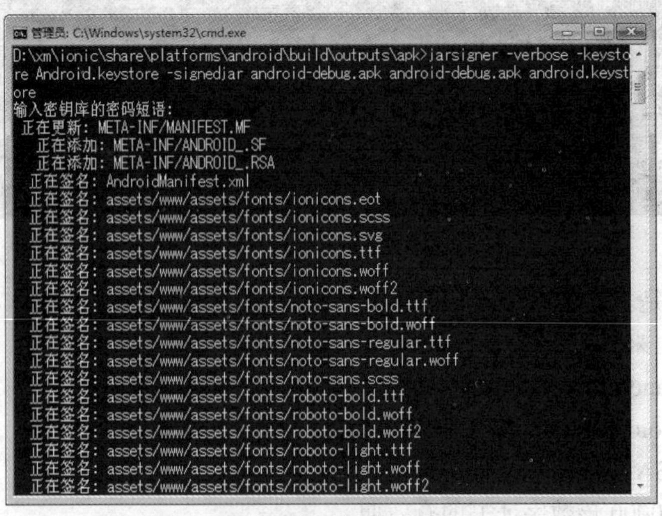

图 8.5　密钥生成效果图

3 优化 APK 文件

优化 APK 文件主要是对其进行压缩,以优化运行时的性能,减少它在设备上占用的空间和内存。需要用到 Zipalign 工具,它是一个 Android 平台上整理 APK 文件的工具,能够优化打包的 Android 应用程序,使 Android 操作系统与应用程序之间的交互作用更有效率,让应用程序和整个系统运行得更快。优化步骤如下所示。

第一步:在 SDK 安装目录下的 build-tools 里找到 zipalign.exe,本书 zipalign.exe 所在目录为 D:\sdk\build-tools\25.0.0。

第二步:把需要优化的 .apk 文件复制到 D:\sdk\build-tools\25.0.0 目录下。

第三步:在该目录下使用"shift 键 + 鼠标右击",选择在此处打开命令窗口,执行以下命令,优化成功后效果如图 8.6 所示。

```
zipalign -v 4 android-debug.apk xing.apk
```

注:android-debug.apk 是优化前的 APK 名字;xing.apk 是优化后的 APK 名字。

图 8.6 优化文件效果图

4 发布 Android 应用

由于国内安卓市场的竞争和碎片化,并没有一个独立的 APP 应用市场来发布 Android 应用,不同的应用市场上传也不同的,以安卓市场的上传为例。

(1)打开安卓市场官网(http://APK.hiAPK.com/),在右上角找到注册按钮,先注册账号成为开发者。如图 8.7 和图 8.8 所示。

(2)账号注册完以后进入网站,点击上传按钮,上传 APP,如图 8.9 所示。

图 8.7 安卓市场官网

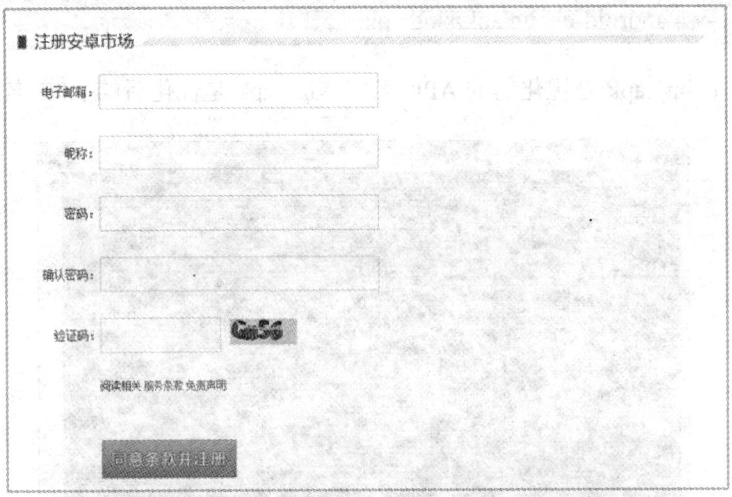

图 8.8 注册界面

图 8.9 上传 APP 效果图

（3）填写应用信息。用户通过这些信息，了解 APP 并决定是否下载。包括应用程序的名称、语言、宣传文字等。如图 8.10 所示。

图 8.10 上传 APP 界面

提示 在书中我们已经实现使用 Ionic 开发 Dance 时代应用。扫描图中二维码,让我们了解更多美观的 Ionic 相关案例。

技能点 2 生成发布 IOS 平台的应用

1 软硬件环境准备

(1)申请苹果开发者账号

打包成 IOS 首先需要加入苹果开发者项目,加入途径详见苹果的官方文档(https://developer.apple.com/programs/)。(打包必须在 Mac OS(苹果系统)上进行(虚拟机和黑苹果一般都比较卡。)

(2)下载安装 Xcode

在 Mac 开发机上打开 Xcode,在菜单中选择 Preferences → Accounts,点击左下角加号,根据提示添加账户信息。如图 8.11 所示。

注意:Mac OS 系统版本要和 Xcode 版本号匹配,也就是旧版本 Mac OS 系统是不能安装新 Xcode 的,如果要在 iPhone 上调试 APP,那 Xcode 版本号就要和 iPhone 版本号匹配,所以最好使用最新版本进行调试和编写,建议使用 Xcode8,本文也将使用 Xcode8 打包。

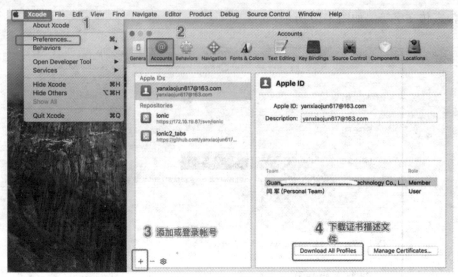

图 8.11 添加账户信息

2 注册 APP ID

APP ID 是发布应用程序的唯一标识。给 IOS 应用程序申请一个 APP ID，首先需要登录苹果开发成员中心（https://developer.APPle.com/membercenter），点击"Account"进入登录界面，如图 8.12 所示。

图 8.12 登录界面

点击证书、ID 及配件文件，进入设置。为应用分配一个 APP ID，如图 8.13 所示。

在"IOS Certificates"页面"Identifiers"下选择"APP IDs"，可查看到已申请的所有 APP ID，点击右上角的加号可创建新"APP ID"，如图 8.14 所示。

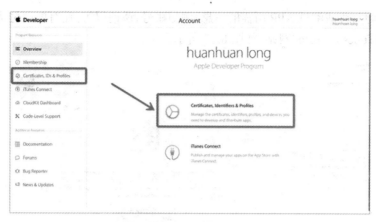

图 8.13　设置界面

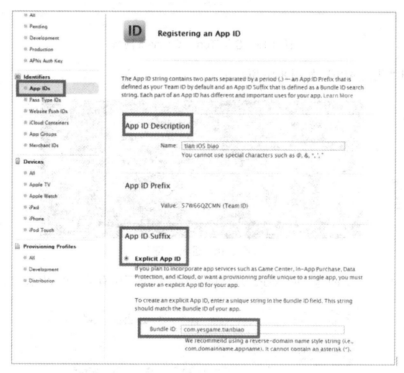

图 8.14　填写信息

● Name，用来描述你的 APP ID，最好是项目名称，方便辨识（不允许中文）。
● Bundle ID（APP ID Suffix），是指应用的代码标识，可以用公司名、应用名来标识。用来标示 APP，使它有一个固定的身份。填写的格式为：com.company.APPName。

3　上架登记

应用上架申请需要使用开发人员的账户信息登录苹果的 iTunes Connect 网址（https://itunes-connect.apple.com），它是苹果公司给个人或企业提供管理自己 APP 的一个平台。在这个平台上开发者可以新建删除和管理自己的 APP 应用，开发者可以根据需求对 APP 应用进行上架与下架，编辑 APP 信息，生成测试 APP 所需的信息，例如账号、邀请码等，还有就内付费等功能。

（1）登录 iTunes Connect 网站，如图 8.15 所示。

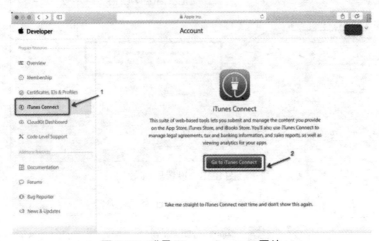

图 8.15　登录 iTunes Connect 网站

（2）点击"我的 APP"，如图 8.16 所示。

图 8.16　进入 APP

（3）点击"加号"，创建新的 APP。这里能看到您账号下所有的 APP，如图 8.17 所示。

图 8.17　创建 APP

（4）进入 APP 的选项信息界面，填写信息，如图 8.18 所示。

图 8.18　填写 APP 信息

设置好相关的信息，类别，价格与销售服务之类的，如果想审核成功后由自己控制发布时间就选择手动发布，默认是自动发布的。APP 描述不要写一些与应用无关的内容，否则易审核不通过，如图 8.19 所示。

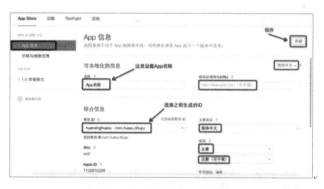

图 8.19　填写 APP 信息

4　生成发布版应用

在项目目录下运行以下命令。

```
ionic build –release ios
```

5　创建应用发布文档

在 Xcode 界面里，在菜单中选择"Product-Scheme-Edit Scheme"，如图 8.20 所示。

图 8.20 Xcode 界面

在弹出框的左栏列表中选择"Archive"选项，确认右侧的"Build Configuration"输入栏选中的是"Release"，然后关闭弹出框，如图 8.21 所示。

图 8.21 更改 Release

如图 8.22 所示选择 Archive 进行打包。

图 8.22 Archive 打包

如果 Archive 不可点击,如图 8.23 所示,请把设备选择为 Generic iOS Device。

图 8.23　Generic iOS Device 打包

出现图 8.24,说明打包已经成功,然后可以上传应用到 APP Store。

图 8.24　打包后

点击右侧蓝色的按钮,依照提示进行则能顺利完成生成应用包的上传工作。回到 Itunes-connect 网站,在 APP 信息中查看一下,是否有构建的版本。如图 8.25 所示。

图 8.25　查看构建版本

然后这个构建版本这里就可以添加代码(注意:如果被拒重新提交,需要在这个地方把上一个构建的版本删除,鼠标放到版本上,删除按钮在版本后面,添加最新构建的版本),如图 8.26 所示。

图 8.26 添加代码

点击加号之后选择代码版本,效果如图 8.27 所示。

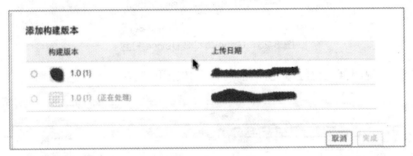

图 8.27 选择代码版本

然后所有东西都填写好了之后,点击页面右上角提交以供审核,然后应用就进入排队等待,效果如图 8.28 所示。

图 8.28 等待发布

　　提示　在了解 Android 签名的操作后,你会不会也想知道 IOS 签名的操作流程。扫描图中二维码,让我们了解 IOS App 签名。

项目八 "Dancer 时代"发布   293

通过下面十四个步骤的操作,实现图 8.2 所示的"Dancer 时代"功能界面及发布该项目。

第一步:设置界面的制作并对界面实现美化。界面中的内容区域是由列表组成。界面代码如 CORE0801 所示。美化后效果如图 8.29 所示。

代码 CORE0801　　设置界面 html

```html
<ion-header>
 <ion-navbar>
  <ion-title>set</ion-title>
  <!-- 头部 -->
 </ion-navbar>
</ion-header>
<ion-content class="set-content">
 <!-- 版本信息 -->
 <div class="set-div">
  <ion-item class="set-item" (click)="d()">
   <span> 版本信息 </span>
  </ion-item>
  <!-- 全屏显示 -->
  <ion-item class="set-item" (click)="a()">
   <span> 全屏显示 </span>
  </ion-item>
  <!-- 屏幕亮度 -->
  <ion-item class="set-item" (click)="b()">
   <span> 屏幕亮度 </span>
  </ion-item>
  <ion-item class="set-item hidd" hidden>
   <div>
    <input type="range" max="1" min="0" step="0.1" value="0" [(ngModel)]="num" (change)="ss()" >
    <p class="spss"></p>
   </div>
  </ion-item>
  <!-- 手电筒 -->
  <ion-item class="set-item" (click)="c()">
   <span> 手电筒 </span>
  </ion-item>
  <!-- 指纹功能 -->
  <ion-item class="set-item" (click)="f()">
```

```
        <span> 指纹 </span>
      </ion-item>
    </div>
  </ion-content>
```

图 8.29　设置界面设置样式后

第二步：创建版本信息界面并配置。

第三步：设置界面跳转，当点击版本信息时跳转到版本信息界面。版本信息插件的相关代码如 CORE0802 所示。

代码 CORE0802　命令窗口安装命令

```
$ cordova plugin add cordova-plugin-app-version
$ npm install --save @Ionic-native/app-version
```

配置及引用代码如 CORE0803 所示。

代码 CORE0803　配置及引用代码

```
import { AppVersion } from '@Ionic-native/app-version';
constructor(private appVersion: AppVersion) { }
this.appVersion.getAppName();
this.appVersion.getPackageName();
this.appVersion.getVersionCode();
this.appVersion.getVersionNumber();
```

第四步：跳转功能的添加，代码如 CORE0804 所示。

代码 CORE0804　设置界面 ts 跳转代码

```ts
import { Component } from '@angular/core';
import { NavController, NavParams } from 'Ionic-angular';
import { AppVersion } from '@Ionic-native/app-version';
import { VersionPage } from '../version/version';
@Component({
  selector: 'page-set',
  templateUrl: 'set.html',
})
export class SetPage {
  constructor(public navCtrl: NavController, public navParams: NavParams,private appVersion: AppVersion) {
  }
  ionViewDidLoad() {
    console.log('ionViewDidLoad Set');
  }
  s1:any;
  s2:any;
  s3:any;
  s4:any;
  d(){
      this.appVersion.getAppName().then((imageData) => {
       this.s1=imageData;
     }, (err) => {
       // Handle error
     });
     this.appVersion.getPackageName().then((imageData) => {
       this.s2=imageData;
     }, (err) => {
       // Handle error
     });
     this.appVersion.getVersionCode().then((imageData) => {
       this.s3=imageData;
     }, (err) => {
       // Handle error
     });
```

```
    this.appVersion.getVersionNumber().then((imageData) => {
      this.s4=imageData;
    }, (err) => {
      // Handle error
    });
    if(this.s1=="undefined"){
      return this.d();
    }else {
      this.navCtrl.push(VersionPage,{ia:this.s1,ib:this.s2,ic:this.s3,id:this.s4})
    }
  }
}
```

第五步：进行版本信息界面制作。该界面由列表组成，部分代码如 CORE0805。设置样式前效果如图 8.30 所示。

代码 CORE0805　　版本信息 html 代码

```html
<ion-header>
  <ion-navbar>
    <ion-title>version</ion-title>
  </ion-navbar>
</ion-header>
<ion-content padding class="verdion-content">
  <div class="verdion-div">
    <ion-item class="verdion-item">
      <span> 名称 </span>
      <i class="verdion-i"></i>
    </ion-item>
    <ion-item class="verdion-item">
      <span> 包名称 </span>
      <i class="verdion-i"></i>
    </ion-item>
    <ion-item class="verdion-item">
      <span> 构建标识符 </span>
      <i class="verdion-i"></i>
    </ion-item>
    <ion-item class="verdion-item">
      <span> 版本 </span>
```

```
        <i class="verdion-i"></i>
      </ion-item>
    </div>
</ion-content>
```

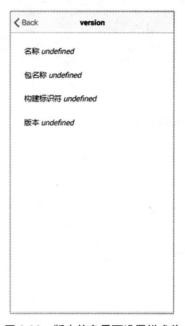

图 8.30　版本信息界面设置样式前

设置版本信息界面样式，需要调整列表样式。部分代码如 CORE0806。设置样式后效果如图 8.31 所示。

代码 CORE0806　SCSS 代码

```scss
.verdion-content{
  background: #FFFFFF;
}
.verdion-div {
  background: #fff;
  margin-top: 5%;
}
.verdion-item {
  padding: 0;
  border-top: 0.55px solid #c8c7cc;
  width: 92%;
  margin-left: 4%;
}
```

```css
.verdion-div .label-ios {
  margin: 0;
}
.verdion-div span {
  display: block;
  float: left;
  line-height: 26px;
  margin-left: 2%;
  font-size: 90%;
  font-weight: 600;
}
.item-md.item-block .item-inner{
  border: none;
}
.verdion-i{
  float: right;
  margin-right: 2%;
  font-size: 80%;
  display: block;
  line-height: 26px;
  color: #6d6565;
}
```

图 8.31　版本信息界面设置样式后

第六步：进行版本信息界面内容的填充，代码如 CORE0807 所示。

代码 CORE0807　版本信息界面 ts 代码

```typescript
import { Component } from '@angular/core';
import { NavController, NavParams } from 'Ionic-angular';
@Component({
  selector: 'page-version',
  templateUrl: 'version.html',
})
export class VersionPage{
  ida:string;
  idb:string;
  idc:string;
  idd:string;
  arr:any;
  constructor(public navCtrl: NavController, public navParams: NavParams) {
    this.ida=navParams.get("ia");
    this.idb=navParams.get("ib");
    this.idc=navParams.get("ic");
    this.idd=navParams.get("id");
    this.arr=[this.ida,this.idb,this.idc,this.idd]
  }
  ionViewDidLoad() {
    for(var i=0;i<this.arr.length;i++){
      document.querySelectorAll(".verdion-i")[i].innerHTML=this.arr[i];
    }
  }
}
```

第七步：安装全屏显示插件，当点击全屏显示时，软件全屏显示。插件的相关代码如 CORE0808 所示。

代码 CORE0808　命令行安装命令

```
ionic cordova plugin add cordova-plugin-fullscreen
npm install --save @Ionic-native/android-full-screen
```

配置及引用代码如 CORE0809 所示。

> **代码 CORE0809　配置及引用代码**
>
> ```
> import { AndroidFullScreen } from '@Ionic-native/android-full-screen';
> constructor(private androidFullScreen: AndroidFullScreen) { }
> this.androidFullScreen.isImmersiveModeSupported()
>   .then(() => this.androidFullScreen.immersiveMode())
>   .catch((error: any) => console.log(error));
> ```

第八步：安装屏幕亮度调控插件，当点击屏幕亮度时，下方出现一个带有进度条的窗口，调整进度条，屏幕亮度随之改变。插件的相关代码如 CORE0810 所示。

> **代码 CORE0810　命令窗口安装命令**
>
> ```
> $ ionic cordova plugin add cordova-plugin-brightness
> $ npm install --save @Ionic-native/brightness
> ```

配置及引用代码如 CORE0811 所示。

> **代码 CORE0811　配置及引用代码**
>
> ```
> import { Brightness } from '@Ionic-native/brightness';
> constructor(private brightness: Brightness) { }
> let brightnessValue: number = 0.8;
> this.brightness.setBrightness(brightnessValue);
> ```

第九步：安装手电筒插件。当点击手电筒时，打开手电筒，再次点击关闭手电筒。插件的相关代码如 CORE0812 所示。

> **代码 CORE0812　命令窗口安装命令**
>
> ```
> ionic cordova plugin add cordova-plugin-flashlight
> npm install --save @Ionic-native/flashlight
> ```

配置及引用代码如 CORE0813 所示。

> **代码 CORE0813　配置及引用代码**
>
> ```
> import { Flashlight } from '@Ionic-native/flashlight';
> constructor(private flashlight: Flashlight) { }
> this.flashlight.switchOn();
> ```

第十步：安装指纹测试插件，当点击指纹时，弹出测试指纹的窗口，当指纹正确时，出现一个对勾图标。插件的相关代码如 CORE0814 所示。

代码 CORE0814 命令窗口安装命令

```
$ ionic cordova plugin add cordova-plugin-fingerprint-aio
$ npm install --save @Ionic-native/fingerprint-aio
```

配置及引用代码如 CORE0815 所示。

代码 CORE0815 配置及引用代码

```typescript
import { FingerprintAIO } from '@Ionic-native/fingerprint-aio';
constructor(private faio: FingerprintAIO) { }
this.faio.show({
    clientId: 'Fingerprint-Demo',
    clientSecret: 'password',
    disableBackup:true
})
.then((result: any) => console.log(result))
.catch((error: any) => console.log(error));
```

第十一步:将功能整合到软件设置界面 ts 中,代码如 CORE0816 所示。

代码 CORE0816 ts 文件

```typescript
import { Component } from '@angular/core';
import { NavController, NavParams } from 'Ionic-angular';
import { AppVersion } from '@Ionic-native/app-version';
import { VersionPage } from '../version/version';
import { AndroidFullScreen } from '@Ionic-native/android-full-screen';
import { Brightness } from '@Ionic-native/brightness';
import { Flashlight } from '@Ionic-native/flashlight';
import { FingerprintAIO } from '@Ionic-native/fingerprint-aio';
@Component({
  selector: 'page-set',
  templateUrl: 'set.html',
})
export class SetPage {
  constructor(public navCtrl: NavController, public navParams: NavParams,private appVersion: AppVersion,private androidFullScreen: AndroidFullScreen,private brightness: Brightness,private flashlight: Flashlight,private faio: FingerprintAIO) {
  }
  ionViewDidLoad() {
    console.log('ionViewDidLoad Set');
```

```
    }
  s1:any;
  s2:any;
  s3:any;
  s4:any;
    // 部分代码省略
    istrue:boolean=true;
  a(){
    if(this.istrue){
      this.androidFullScreen.isImmersiveModeSupported()
        .then(() => this.androidFullScreen.immersiveMode())
        .catch((error: any) => console.log(error));
      this.istrue=false;
      }else {
      this.androidFullScreen.isImmersiveModeSupported()
        .then(() => this.androidFullScreen.showSystemUI())
        .catch((error: any) => console.log(error));
      this.istrue=true;
    }
  }
  isfalse:boolean=true;
  b(){
    if(this.isfalse){
      document.querySelectorAll(".hidd")[0].removeAttribute("hidden");
      this.isfalse=false;
    }else {
      document.querySelectorAll(".hidd")[0].setAttribute("hidden","hidden");
      this.isfalse=true;
    }
  }
  num:number;
  ss(){
    var numb=this.num;
    var span=document.querySelectorAll(".spss")[0];
    span.innerHTML=""+numb+"";
    this.brightness.setBrightness(numb);
  }
  c(){
```

项目八 "Dancer 时代"发布

```
    this.flashlight.toggle();
}
f(){
    this.faio.show({
        clientId: 'Fingerprint-Demo',
        clientSecret: 'password', //Only necessary for Android
        disableBackup:true  //Only for Android(optional)
    })
    .then((result: any) => console.log(result))
    .catch((error: any) => console.log(error));
}
}
```

第十二步：项目的功能及结构已经编写完成通过下列命令进行打包测试。

```
ionic build android --release
```

第十三步：通过下列命令生成秘钥。

```
keytool -genkey -v -keystore android.keystore -alias android.keystore -keyalg RSA -keysize 2048 -validity 10000
```

第十四步：正式发布 APK。

```
jarsigner -verbose -keystore Android.keystore -signedjar enhanced_signed.apk enhanced.apk android.keystore
```

至此，"Dancer 时代"功能及发布该项目。

本项目通过发布"Dancer 时代"项目，对 Ionic 项目在 Android 及 IOS 手机下的发布流程具有初步了解，并详细了解了 Ionic 环境如何签名和优化 APK，通过 Ionic 项目发布成功，具有根据需求分析和设计制作并发布项目的能力。

| release | 发布 | keystore | 秘钥 |
| validity | 有效的 | certificate | 证书 |

verbose	详细的	registered	注册
account	账号	debugging	调试
identifiers	标识符		

一、选择题

1. 打包 APK 时使用命令生成签名文件，其中 -v 表示（　　）。
   A. 工具名称　　　　　　　　B. 将生成证书的详细信息打印出来
   C. 执行的是生成数字证书操作　D. 生成密钥文件所采用的算法

2. 对 APK 文件签名需要使用 JDK 附带的（　　）工具为 APK 签名。
   A.jarsigner　　　B.keystore　　　C.RSA　　　D.keytool

3. 优化 APK 需要用的工具是（　　）。
   A.jarsigner　　　B.Zipalign　　　C.keystore　　　D.keytool

4. 下列语句及意思错误的是（　　）。
   A.-keystore android.keystore 证书的文件名
   B.-alias android.keystore 表示证书的别名
   C.-keyalg RSA 生成密钥文件所采用的算法
   D.-validity 10000 该数字证书的大小

5. 下列语句中，表示签名后生成的 APK 名称的是（　　）。
   A.-signedjar enhanced_signed.apk　　　B.-keystore android.keystore
   C.enhanced.apk　　　　　　　　　　　D.android.keystore

二、填空题

1. Ionic 功能开发完成后，就需要将应用打包发布到应用市场。有两种打包的版本：一种是 debug 调试版，另一种是 _____ 。

2. 打包 debug 调试版需在工程目录下执行 _____ 。

3. 生成签名的秘钥需要使用 JDK 附带的 _____ 工具，该文件一般在 JDK 路径下的 bin 目录下。

4. 优化 APK 文件是对其进行压缩，以优化运行时的性能，减少他在设备上占用的空间和内存。需要用到 _____ 工具。

5. Bundle ID (APP ID Suffix) 是指应用的 _____ ，可以用公司名、应用名来标识。用来标识 APP，使它有一个固定的身份。

三、上机题

使用 Ionic 知识实现下列要求的效果。要求：
1. 添加屏幕亮度插件；

2. 使用 JavaScript 进行滑动条的拖动来改变亮度,效果如下图。